Fatigue of Engineering Plastics

Fatigue of Engineering Plastics

RICHARD W. HERTZBERG
JOHN A. MANSON

Materials Research Center
Lehigh University
Coxe Laboratory
Bethlehem, Pennsylvania

 1980

ACADEMIC PRESS

A Subsidiary of Harcourt Brace Jovanovich, Publishers

New York   London   Toronto   Sydney   San Francisco

ACADEMIC PRESS, INC.
111 Fifth Avenue, New York, New York 10003

United Kingdom Edition published by
ACADEMIC PRESS, INC. (LONDON) LTD.
24/28 Oval Road, London NW1   7DX

Library of Congress Cataloging in Publication Data

Hertzberg, Richard W.   Date
   Fatigue of engineering plastics.

   Includes bibliographical references and index.
   1. Plastics––Fatigue. I. Manson, John A.,
Date   joint author. II. Title.
TA455.P5H45   620.1'9233   79–6786
ISBN 0–12–343550–1

PRINTED IN THE UNITED STATES OF AMERICA

80 81 82 83   9 8 7 6 5 4 3 2 1

D
620 · 1923'3

HER

To our wives, Linda and Tally, and our children, Michelle, Jason, Patricia, John, William, and Barbara.

Contents

3   Fatigue Crack Propagation

4   Fatigue Fracture Micromechanisms in Engineering Plastics

5   Composite Systems

Preface

This book was written for polymer scientists and engineers who seek a more complete understanding of fatigue processes in engineering plastics as affected by polymer chemistry, architecture, and processing. It should also be of interest to materials and design engineers who utilize such materials for applications in which superior fatigue performance is demanded. The book will serve as a useful reference for workers in the field and for advanced students. While some aspects of fatigue have been treated in a few reviews and book chapters, these have been designed for more limited audiences. We believe that "Fatigue of Engineering Plastics" will provide the first integrated treatment of polymer structure, morphology, and continuum aspects of fatigue, as well as the micromechanisms of the fatigue process.

In view of increased use of plastics in components and structures where cyclic loads are experienced, a reference work on the subject of fatigue of engineering plastics is well justified. For example, the extensive use of plastics in matrixes of composite materials and in the manufacture of prosthetic devices, machine components, and piping demands that the overall mechanical response of these materials be characterized more fully. The fact that field failures related to polymer fatigue mechanisms are being reported further emphasizes the need for understanding polymeric response to cyclic loads.

At the current stage of development of both testing and research on fatigue in polymers, it seemed appropriate to review the general question of fatigue in engineering plastics and composites. In particular, it was thought desirable to emphasize the roles played by testing and material parameters, the distinction between failure by thermal softening and by the propagation of a flaw, and the use of fractography to infer the micromechanisms of failure and the applicability of fracture mechanics. Whenever possible, phenomenological and continuum-mechanical aspects are brought together, and the chemistry is related to the mechanics. The discussion includes reference to the prediction of fatigue life.

A second major aim of this monograph is to bring together bodies of literature and approaches that have tended to be restricted to particular cases and disciplines. Thus, regrettably, the typical chemist is unfamiliar with fracture phenomena, and the typical fracture mechanics expert is unfamiliar with molecular structure and properties. Also, although fatigue processes in polymers exhibit some characteristics seen in metals, several profound differences exist and must be taken into account. With respect to scope, while engineering plastics (as a general class of polymer) and conventional polymer composites are emphasized, limited attention is given to related systems such as adhesive joints and polymer–concrete systems.

"Fatigue of Engineering Plastics" begins with an overview of polymer fatigue as it relates to deformation and fracture processes. Fatigue damage under cyclic load and cyclic strain conditions is then described with particular attention being given to variables that affect the onset of hysteretic heating. An analysis of fatigue crack propagation (FCP) in terms of fracture mechanics concepts follows and focuses on material and test variables that strongly influence FCP response. Fatigue fracture micromechanisms are then analyzed with the aid of electron microscopic observations and provide the reader with valuable information regarding the postmortem examination of fracture surfaces of polymer components that have failed in service. The final segment of the book considers the fatigue behavior of polymer composites that consist of polymeric matrixes with both fibrous and particulate second phase additions.

Lehigh University
June, 1980

RICHARD W. HERTZBERG
JOHN A. MANSON

Acknowledgments

During the past twelve years, our joint research program on polymer fatigue has received support and encouragement from numerous professional colleagues. We recall critical discussions with Drs. A. Peterlin, R. F. Boyer, E. H. Andrews, P. W. Beardmore, F. X. de Charentenay, J. M. Liégeois, E. A. Collins, L. J. Broutman, J. A. Sauer, H. H. Kausch, J. G. Williams, R. P. Kambour, R. S. Porter, and V. Tamuzh at Lehigh University and elsewhere, which impacted greatly on our research. Without the significant contributions from our colleagues, Drs. S. L. Kim, M. D. Skibo and P. E. Bretz, and students, C. M. Rimnac, S. M. Webler, M. T. Hahn, A. Ramirez, R. Lang, S. Qureshi, J. Janiszewski, and J. Michel, our research and the preparation of this manuscript would have been severely restricted.

We wish to thank the Air Force Materials Laboratory, the Pennsylvania Science and Engineering Foundation, the Army Research Office-Durham, National Science Foundation, the Office of Naval Research, the Materials Research Center, Lehigh University, and E. I. du Pont de Nemours & Co. for financial support of our polymer fatigue research program. Also, Dr. W. Kurz and the École Polytechnique Fédérale de Lausanne, Switzerland, kindly provided an ideal atmosphere during the summer of 1976, which enabled us to conceive the plan and scope of this monograph. We appreciate the assistance and cooperation of our Lehigh University colleagues in the

Materials Research Center, the Department of Metallurgy and Materials Engineering, and the Department of Chemistry, and in particular, the secretarial assistance of L. Valkenburg, J. Svirzofsky, and J. Loosbrock. Finally, the enduring patience of our wives and children is greatly appreciated. Their understanding and support during the book's gestation period will always be remembered.

List of Acronyms

ABS	acrylonitrile–butadiene–styrene copolymer	N6	nylon 6
		N66	nylon 66
A/E	amine/epoxy ratio	PA	polyacetal
bfrp	boron-fiber-reinforced plastic	PC	polycarbonate (or, in the field of concrete materials, polymer concrete)
cfrp	carbon-fiber-reinforced plastic		
CLPS	cross-linked polystyrene		
CT	compact tension	PE	polyethylene
DGB	discontinuous growth band	PIC	polymer-impregnated concrete
DOP	dioctyl phthalate	PMMA	poly(methyl methacrylate)
FCP	fatigue crack propagation	POM	polyoxymethylene
FSF	frequency sensitivity factor	PP	polypropylene
gfrp	glass-fiber-reinforced plastic	PPCC	polymer-portland-cement-concrete
HI-	high-impact-		
HIPS	high impact polystyrene	PPO	poly(phenylene oxide)
ksi	one thousand pounds per square inch	PS	polystyrene
		PSF	polysulfone
LDPE	low-density polyethylene	psi	pounds per square inch
M	molecular weight	PVC	poly(vinyl chloride)
MBS	methacrylate-butadiene-styrene	PVDF	poly(vinylidene fluoride)
MPa	megapascal $\equiv$ meganewton per square meter	RH	relative humidity
		SEM	scanning electron microscope
$\text{MPa} \cdot \text{m}^{1/2}$	megapascal times square root meter	SEN	single edge notched
		S–N	stress-cycle life
MWD	molecular weight distribution	TEM	transmission electron microscope
nBA	n-butyl acrylate		

List of Symbols

a = crack length

β = beta damping peak

χ = displacement amplitude or maximum static strength

da/dN = fatigue crack growth rate

δ = crack opening displacement or phase angle between stress and strain

D = damping

E = elastic modulus

E_d = dynamic elastic modulus

γ = surface energy

G = shear modulus

$\mathscr{G} = dU/da$ = strain energy release rate

$I(t)$ = creep compliance

J'' = loss compliance

K = stress intensity factor

K_c = fracture toughness

K_{max} = maximum stress intensity factor

K_{min} = minimum stress intensity factor

ΔK = stress intensity factor range

ΔK_{th} = threshold stress intensity range

ΔK^* = ΔK for da/dN = const

$\lambda = K_{max}^2 - K_{min}^2$

M_c = critical value of M to develop tensile strength

M_n = number-average molecular weight

M_v = viscosity-average molecular weight

M_w = weight-average molecular weight

M_z = z-average molecular weight

N = number of cycles

N^* = cyclic stability of DGB

v = frequency

P = probability

Q_a = endothermal peak

$R = \sigma_{min}/\sigma_{max}$

r_y = plastic zone radius

$\sigma_a = \frac{1}{2}(\sigma_{max} - \sigma_{min})$

$\sigma_{avg} = \sigma_{11} + \sigma_{22} + \sigma_{33}$ = hydrostatic pressure

σ_b = breaking stress

σ_{ys} = yield strength

$\sigma_{11}, \sigma_{22}, \sigma_{33}$ = principal stress

τ = life

T = tear energy density

U = strain energy density

U_0 = activation energy

Fatigue of Engineering Plastics

1 | *Introduction to Fatigue*

1.1 GENERAL ASPECTS

Fatigue, the loss of strength or other important property as a result of stressing over a period of time, is a very general phenomenon in most materials and even living organisms.* While the response to a monotonic stress, as in static creep rupture, may be considered as a limiting case, the term "fatigue" usually refers to the more general case of intermittent stressing. Various atomic or molecular processes (often unknown in detail) may take place during the fatigue phenomenon, some beneficial and some deleterious. However, with a few exceptions, the balance between deterioration and strengthening is eventually struck in favor of deterioration and failure, usually as a result of stresses that are small in comparison to those required for failure due to a monotonic stress.

Thus, as shown in Fig. 1.1, a typical material subjected to an intermittent stress (typically cyclic or pulsed) will ultimately fail at progressively shorter times, the higher the applied stress. In some cases there may also be a limiting

---

* Although mental and body fatigue is all too familiar to us, it should not be supposed that the term is yet another example of barbarous jargon. Indeed the technical use has a long and respectable history, going back at least to the 1700s [1]. The term itself is derived from *fatigare* (Latin) meaning "to tire." The root *fati* refers literally to an opening or yawning phenomenon vividly descriptive of fatigue in people and, especially, in view of the ubiquity of crazing and cavitation, in materials as well.

1

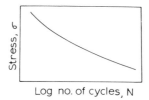

Fig. 1.1   Typical fatigue response in a polymer: the relationship between applied stress σ and number of cycles to failure N. Strain may also be plotted instead of stress.

low stress (the "endurance limit") below which failure does not occur on any realistic time scale. It should also be noted that failure may be defined not only in terms of fracture, but also in terms of the loss of functional effectiveness as defined by failure to meet some design criterion such as minimum strength or stiffness.

The competitive processes leading to fatigue failure in polymers and their correlation with loading conditions, material structure, and composition are (or should be) of great interest to the scientist. The time dependence of behavior that is characteristic of polymers as a class, but rare in other materials, poses a special challenge. However, as polymers and their composites are being used increasingly for load-bearing purposes under fatigue conditions, the engineer responsible for component design or materials selection must also know the fatigue characteristics and have some understanding of the effects of loading and material variables. This is especially so because of the insidious nature of fatigue, and the complexities of the structure of polymers compared to metals. Unfortunately, in accordance with a corollary of Murphy's law, failure of an important component is bound to occur at a time and place that is at least inconvenient and costly, and possibly dangerous. While failures in toys and other common artifacts may pose relatively simple problems, failure of components in, for example, vehicles, aircraft, process equipment, and piping, may have consequences that are far from trivial. The problem is compounded by the facts that some toughened plastics may be sensitive to some fatigue conditions [2], and that the existence of fatigue loading in service is not always recognized. For example, pressure pulses in plastic pipes may cause premature brittle failure in a normally ductile polymer at surprisingly low loads, and premature failure of a plastic boat hull in fatigue has also been reported [3]. As concern for product liability on the part of designers and manufacturers increases, so must concern for the possibility of fatigue failure and for designing against it.

In materials selection and component design, the empirical characterization of fatigue in polymers is undoubtedly useful and necessary. However, sophistication in these functions requires more fundamental understanding of the phenomena underlying the overall fatigue process. While curves of the type shown in Fig. 1.1 (the traditional "S–N" curve depicting stress versus the number of cycles to failure) can be used to select materials, they tell nothing about such phenomena as the initiation and propagation of crazes or cracks, yielding and drawing, or the storage, release, or dissipation of energy.

As mentioned above, the question of fatigue in materials is not new; fatigue has been recognized as a problem for at least about 200 yr. In the days of sailing vessels, the term was applied to the failure of masts caused by hoisting too many sails too often [1]. Later, when the steam engine revolutionized transportation, fatigue in rolling-stock axles became a serious problem. Thus the stage was set for the scientific and engineering study of fatigue in metals—a study that has continued since 1871 [4], and that is now well advanced. A particularly strong impetus to research on fatigue in metals has been given by the advent of steel ships, oil pipelines, and jet aircraft, and the use in design of concepts of fracture mechanics is common and often obligatory [5].

However, in spite of the need to understand fatigue in polymers, the state-of-the-art is not nearly as well advanced as is the case with metals. With some exceptions, the phenomenon of fatigue in polymers and composites has received little fundamental attention, and not as much engineering attention as is warranted. To be sure, a great deal of empirical work has been done, but until recently fundamental and well-designed scientific and engineering studies have been restricted to rubbers and fibers. Thus in 1971, it was stated that fatigue in polymers constitutes one of the unsolved problems in polymer science [6]. Fortunately, our knowledge has advanced significantly since then, though many unsolved problems still exist.

The understanding of fatigue in polymers has been hindered by three major problems. First, the continuum-mechanical approach to fracture developed by the physicist or mechanician does not in itself describe the atomic and molecular processes familiar to the chemist. Thus, two fundamentally different views of matter confront each other and require correlation and reconciliation. Second, the ultimate phenomenon of fracture involves gross, irreversible, and nonlinear deformations, in contrast to the small-scale, reversible, and linear deformations often measured by the physical chemist in order to characterize the effects of molecular properties and polymer composition. Third, the precise nature of the competitive processes that are balanced out in fatigue, and that give rise to the especially damaging effect of intermittent loading, is not known in detail. Nevertheless, if one is willing to tolerate ambiguity, to accept the duality of continuum and molecular notions as representing different aspects of truth, and to see how far one can apply linear descriptions to nonlinear systems (or appropriately modify the descriptions), progress can be made. Indeed the use of a continuum-fracture-mechanics approach has proved to be useful in characterizing fracture in metals; complementary approaches based on consideration of either applied stress or fracture energy have also been developed [5]. With fracture in polymers, which exhibit viscoelasticity and nonlinearity in behavior, even a linear-elastic-fracture-mechanics approach has shown considerable promise in at least correlating behavior, and nonlinear models are also being developed [7, 8].

At the current stage of development of both testing and research on fatigue in polymers, it seemed appropriate to review the general question of fatigue in engineering plastics and composites. In particular, it was thought desirable to emphasize the roles played by testing and material parameters, the distinction between failure by thermal softening and by the propagation of a flaw, and the use of fractography to infer the micromechanisms of failure and the applicability of fracture mechanics. Whenever possible, phenomenological and continuum-mechanical aspects are brought together, and the chemistry related to the mechanics. Implications with respect to the prediction of fatigue life are discussed, and directions for needed research will become evident.

A second major aim of this monograph is to bring together bodies of literature and approaches that have tended to be restricted to particular cases and disciplines. Thus, regrettably, the typical chemist is unfamiliar with fracture phenomena, and the typical fracture mechanics expert is unfamiliar with molecular structure and properties. Also, although fatigue processes in polymers exhibit some characteristics seen in metals, several profound differences exist, and must be taken into account. With respect to scope, while engineering plastics (as a general class of polymer) and conventional polymer composites are emphasized, limited attention is given to related systems such as adhesive joints and polymer-concrete systems. However, fibers and elastomers fall outside the scope of this monograph.

1.2 PREVIOUS LITERATURE

Any monograph is bound to reflect the personal views of its authors, and must necessarily be selective in its discussion of the literature. For other views and for completeness, it is useful to consider other publications on this subject.

The earliest comprehensive review on fatigue in polymers was published by Dillon [8a] in 1950. Although this review predated such developments as the time–temperature superposition principle and the fracture mechanics approach, the state of knowledge was evaluated quite critically, phenomenological aspects were treated thoroughly, and an excellent outline of fatigue testing was presented.

Following Dillon's review, little comprehensive literature on fatigue in polymers and related topics was published until a lengthy and detailed monograph edited by Rosen [9] presented theoretical and phenomenological aspects of the kinetics, energetics, and morphology of polymer fracture. In particular, Wolock and Newman [10] and Landel and Fedors [11] discussed fracture surface topography and failure in amorphous polymers, respectively. Following an overview of fracture (including fatigue) in polymers [12], Andrews published a clear exposition of his views on static and fatigue failure in a concise book [13]. In later reviews, Andrews also described the distinction between creep, thermal, and mechanical aspects of fatigue (a distinction not always recognized) and outlined fracture mechanics approaches to the characterization of fatigue [14, 15]. Some effects of composition were also discussed by Bucknall *et al.* [16]. An interesting discussion of fatigue is given in a chapter on fracture by Vincent [17], and in a review by Hearle of fatigue in polymers, especially fibers [18]. Two Russian reviews by Regel and Lekovskii [19] and Tamuzh [20] also appeared at about the same time.

In 1972 Plumbridge [21] treated the fatigue of polymers and metals together; this attempt of generalization is useful in view of the need to bring together different kinds of specialists. Shortly thereafter, in a precursor to this volume, the authors critically reviewed the field of fatigue in polymers, with emphasis on the role of molecular structure and composition, and micromechanisms [22]. More recent reviews by Beardmore and Rabinowitz [23] and by Schultz [24] emphasize the phenomenological aspects and micromechanisms involved. General discussions of fracture in multicomponent systems are provided by Bucknall [25, 25a], while Owen [26], and Harris [27] emphasize fatigue in fibrous composites. Recent critical reviews of the molecular aspects of fracture and of current fracture mechanics approaches are provided by Andrews and Reed [28], and Williams [8], respectively; these topics are also discussed in detail by Kausch [29].

Relevant aspects of deformation behavior and micromechanisms are discussed in detail by Berry [30], Kambour and Robertson [31], Ward [32, 33], and Bucknall [25, Chaps. 5–10]. Although elastomers are excluded from significant discussion in this book, the reader who wishes details on their fatigue behavior is referred to papers and reviews by Beatty [34], Lake and Lindley [35], Lake [36], Gent [37], and Eirich and Smith [38]. The topic was also discussed in our earlier critical review [22].

Apart from these general references cited, the organized and readily available literature in English on fundamental aspects is not well developed. Not many textbooks and monographs discuss fatigue, though there are brief references by Billmeyer [39], Rodriguez [40], and Nielsen [41]. (Curiously, in a contemporary design guide for plastics [42] fatigue is not explicitly discussed.) Some engineering data (S–N curves) are available from manufacturers, some in the literature [43, 44], and some in reports (especially on composites); where possible, references cited in this monograph are restricted to reviewed publications.

1.3   GENERAL CONSIDERATIONS

Before discussing specific examples of fatigue response, a brief review of related phenomena and parameters is in order. This will provide a background against which to view accounts of experiments and theoretical developments. Further details may be found in the references cited in Section 1.2 and in the following chapters.

As will be seen, fatigue loading superimposes additional material responses and consequences on those commonly encountered in "static" deformation sequences such as stressing in simple tension. Awareness of the multiplicity of responses and their frequently competitive nature is clearly important both in the design of fatigue experiments and in the interpretation of results obtained.

In any experiment on material behavior, it is assumed that the contributions of the material per se are revealed as distinct from the contributions of the test per se. Also, in engineering research the variables selected should reflect as much as possible the variables that are important in the actual situation being modeled. Unfortunately, even with conscious attention to the design of experiments, it is not always easy to achieve these aims, and the effects of variables are often confounded, knowingly or unknowingly. Fatigue testing is no exception, and often constitutes a lamentable example of poor experimental planning. Indeed, many fatigue tests and experiments described in the scientific and technological literature are of such a nature that only limited inferences may be drawn [14, 22]. In fairness, it must be recognized, of course, that the needs of engineering designers and scientists do not always coincide.

1.3.1 Stress Systems

Most of this discussion is based upon the chapters by Dillon [8a] and Andrews [14]. As summarized by Andrews, fatigue tests and experiments involve the following:

(1) a periodically varying stress system having a characteristic stress amplitude $\Delta\sigma$;
(2) a corresponding periodic strain amplitude $\Delta\varepsilon$;
(3) a mean stress level σ_m;
(4) a mean deformation ε_m;
(5) a frequency v;
(6) a characteristic wave form (sinusoidal, square, etc.) for both the stress and strain;
(7) ambient and internal temperatures, which in general will not be precisely identical; and
(8) a given specimen geometry, including notches (if any).

Of course, in laboratory tests or service environments, many complex combinations are possible, e.g., a sequence of different stress amplitudes, mean stresses, and frequencies of loading. Interactions between stress fields are also often important, especially with composites.

In practice, tests usually fall into five categories (for examples see the subsequent chapters):

(1) periodic loading between fixed stress limits, in tension or compression;
(2) periodic loading between fixed strain limits, in tension or compression;
(3) reversed bending stress (implicit in flexing a sheet in one dimension);
(4) reversed bending stresses in two dimensions (by rotary deflection of a cylindrical specimen); and
(5) reversed shear stresses (obtained by torsional deformation).

Unfortunately, the multiplicity of loading modes and parameters experienced in service and used in tests often makes it difficult to compare results. For example, while most experiments are conducted under type (1) conditions (in tension), it will be seen that behavior in other modes may be quite different; moreover, many service conditions involve states of combined stress. Thus the needs of fundamental study and of quality control tests may be quite different. In any case, it is always useful to design tests with subsequent interpretation in mind (see discussions in Chapters 2 and 3, and in ref. [14]). Specimen configurations must also be selected with interpretability in mind, in order to be sure that true *material* constants are obtained. Fortunately analytical treatments [5, Chap. 8; 45, 46] make it possible to use a large number of specimen geometries that can include a variety of shapes more characteristic of engineering situations (see also Section 2.3.1).

Once the mode of testing and the test variables are selected, one must consider how best to use the data in order to rank materials in order of fatigue resistance or to predict fatigue life. The classical fatigue test in metals or polymers is exemplified by the S–N curve already discussed (Fig. 1.1), in which the stress range $\Delta\sigma$ (or strain range $\Delta\varepsilon$) is plotted against the number of cycles to failure (on a logarithmic scale). For obvious reasons curves of this type have long been used by the design engineer, who after all may not care to know the mechanism by which failure occurs. If the anticipated stress range (and mean stress) in service is below the horizontal asymptote at low values of $\Delta\sigma$, fatigue failure need not be expected. In other cases, lifetime can often be estimated, providing adequate statistical analysis can be made.

However, as discussed in Chapter 2, tests of this type are not helpful from a fundamental standpoint. The information is specific only to the particular specimen used (which may not be representative of a typical specimen in service) and does not illuminate the mechanism of failure. Specifically, the S–N curve does not in general distinguish between the initiation of cracks and their propagation to catastrophic failure. This is a serious deficiency in terms of both fundamental and practical implications, for if there are adventitious flaws capable of growth under the load applied, crack *propagation* rather than *initiation* may well constitute the major failure process in determining the lifetime of a specimen (see Fig. 1.2, ref. [47], and Chapters 2 and 3).

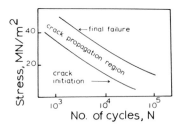

Fig. 1.2 Fatigue response of a polymer showing the relationship between applied stress σ and number of cycles to failure N for both initiation of fatigue cracks and final failure.

1.3.2 Molecular Aspects of Fatigue

When a polymer system is subjected to an applied stress, a great variety of chemical and rheological events (often having opposing effects) may occur in a given polymer at a given temperature. The net effect of the several competing processes depends on many factors, including the temperature, the time scale of the experiment with reference to the time scale of the process, the environment, the basic molecular properties of the polymer, and the composition. Some of the major properties and deformation modes that are important in the fatigue behavior are

(1) molecular characteristics, such as molecular weight (and its distribution) and thermodynamic state;
(2) chemical changes, such as bond breakage;
(3) homogeneous deformation (elastic, anelastic, and viscoelastic);
(4) inhomogeneous deformation (crazing and shear banding);
(5) morphological changes, such as drawing, orientation, and crystallization.
(6) transition phenomena, such as the glass-to-rubber and secondary transitions; and
(7) thermal effects, such as hysteretic heating.

Of course, many interactions between these factors can take place.

1.4 MOLECULAR CHARACTERISTICS OF POLYMERS

Thus deformation and flow in polymers depend on several of the characteristics just mentioned and especially on molecular structure, molecular weight and its distribution, composition, and morphology (e.g., crystalline texture). Some of these factors, namely, the molecular structure and weight, are fixed during synthesis (unless chemical degradation occurs during fabrication); however, the morphology is determined by the thermal and mechanical history experienced by the polymer prior to (and sometimes during) testing ([41], pp. 235, 261). Together with temperature and pressure, all these factors combine to determine the thermodynamic state (characterized by specific volume, enthalpy, and entropy) and the corresponding viscoelastic state (typically characterized by the dynamic modulus or compliance).

The characteristic temperatures for transitions from one state to another (see below), and the values of isothermal viscoelastic parameters, depend on the rate of loading (i.e., on the time scale of the experiment, in accordance with an appropriate relationship between rate and temperature [48]).

At sufficiently low stresses and strains, the mechanical response is reversible (*linear viscoelastic behavior*), while at higher stresses and strains an irreversible component of deformation becomes evident (*nonlinear viscoelastic behavior*). Cyclic loading superimposes on mechanical response the phenomenon of hysteretic heating (Chapter 2), which in turn depends on the viscoelastic state. Of course, if such heat is not entirely dissipated to the surroundings (as is often the case with materials like polymers that have relatively low thermal conductivities), the temperature rises, thus changing the viscoelastic state. Another complication is that, even in nominally homogeneous polymers, the deformation is often inhomogeneous in nature. For example, different degrees of deformation occur in the amorphous and crystalline regions of semicrystalline polymers, and even in amorphous polymers, localized as well as general deformation of the whole specimen may take place. In addition, the textures of amorphous polymers are often far from homogeneous [49]. Thus, during deformation and fracture, a variety of events takes place, including homogeneous shearing, heterogeneous shearing (shear banding), dilatational (crazing) processes, and bond rupture. In view of the large number of interacting factors and their complexities, it is not surprising that their role in fatigue has been incompletely studied; moreover, many of these factors are not encountered in metallurgy or ceramics. Thus it is worthwhile to review briefly some of the major factors and phenomena anticipated in fatigue.

1.4.1 Chemical Effects

When polymer molecules are subjected to high stresses, covalent bonds [e.g., C—C bonds, which usually have a cohesive energy of about 0.3 MJ/mole (80 kcal/mole)] may be broken. This may occur not only as a consequence of crack growth, but also as a consequence of large shearing forces during fabrication [50]. If sharp flaws are present, the nominal stress required for rupture may, of course, be significantly reduced due to the concentration of stress at the tip of the flaw. Indeed, the rupture of bonds plays a central role in theories that predict the lifetime of stressed polymers on the basis of an accumulation of damage. (For detailed discussion, see, for example, Zhurkhov and Thomashevskii [51], the recent reviews by Andrews and Reed [28]) and by Kausch [29], and a brief discussion in Section 2.7).

To what extent is bond rupture important in determining the fracture energy of polymers and the corresponding resistance to fatigue? To be sure, the occurrence of bond rupture has been demonstrated in many polymers (subjected to either static or cyclic loading) by means of electron spin resonance (ESR) techniques. Thus, in an extensive series of elegant experiments by Zhurkov and Thomashevskii [51] and by deVries, Williams *et al.* [51a–55], a close relationship between stress and bond breakage was confirmed and lifetimes predicted in terms of the rates of bond rupture. In fatigue, it was found (Fig. 1.3) that the rate of free-radical production in cyclic deformation was in phase with the load excursions. However, it should be noted that the energy required for fracture is invariably much greater than that required to break the covalent bonds in the plane* ahead of a growing crack or in a region close to the plane (see, e.g., Andrews [4]; Andrews and Reed [28], and the following discussion). In fact, the contribution of bond rupture to the overall fracture energy is significant (and still *relatively* small) only in polymers such as cross-linked or crystalline ones [56] whose restrictions on sequential mobility can permit the transfer of large stresses to many molecules or segments in a zone near the crack plane and sufficiently inhibit segmental motion. Thus in a typical glassy polymer, bond rupture contributes about 0.1 J/m^2 to fracture energies that are on the

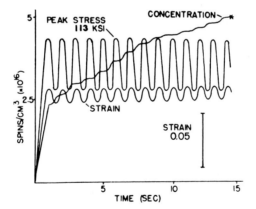

Fig. 1.3 Stress, strain, and unpaired electron concentration versus time for cyclic loading superimposed on a static load. [M. L. Williams and K. L. DeVries, *Proc. 5th Int. Cong. Rheol.* **3,** 139 (1970). Corpyright © by the University of Tokyo Press. Reprinted by permission of the University of Tokyo Press.

* It should be noted that bond breakage is not confined strictly to the plane ahead of the crack. Up to an order-of-magnitude more covalent bonds are broken than are in the formal crack plane; some of these bonds are in planes that are perpendicular to the crack.

order of 10^2–10^3 J/m^2; with crystalline polymers, the contribution of chain rupture is still not greater than about 10% of the total energy.

Bonds may also be broken as a consequence of chemical reactions that lead to the formation of weak bonds. Thus chain scission due to oxidation by oxygen or ozone is important in the static and fatigue fracture of rubbers; the effects of oxidation have been studied extensively by Gent, Lindley, and others [13, 14, 42–44]. While some degree of cross-linking may take place concurrently, the net effect is generally a reduction in modulus and strength. Although one might well expect similar effects with nonrubbery polymers as well, this question does not seem to have been addressed.

Another important chemical effect is *stress cracking* induced by the inter- action between a polymer and a nonsolvent in the presence of a mechanical stress [25, p. 171; 57, 57a]. (Internal residual stresses are often sufficient to induce stress cracking in the absence of an external stress.) Interactions of this kind are notorious precursors of failure in many polymers; in fact, stress cracking continues to be one of the most serious engineering problems with polymers. (One finds similar counterparts of stress and environment assisted cracking in metallic alloys such as in hydrogen embrittlement and stress corrosion cracking.) Many theories have been proposed to explain the phenomenon. One of the most plausible was proposed by Gent [58] who suggested that a nonsolvent, always absorbed to at least some extent, reduced the glass-transition temperature and modulus to the point that cavitation and crack formation can occur. This general notion of plasticization has also received support from experiments by, among others, Kambour *et al.* [59] and Andrews *et al.* [60]. It is interesting that even liquid nitrogen can serve to initiate crazing at very low temperatures [61]. Other environmental studies under both static and dynamic conditions have been reported by Williams *et al.* [57, 57a], and Mai and Williams [62], and the kinetics estab- lished (see Section 3.6).

Other examples of the ability of a liquid to change the state of a polymer include plasticization or antiplasticization by small amounts of water or other reagents [63] and the transformation of polycarbonate from a glassy to a crystalline state under the influence of small quantities of, for example, acetone [64]. To be sure, not all effects of liquids are deleterious; in some cases, the sorption of water may enhance toughness and fatigue resistance (see references [25b, Chap. 12; 65, 66]; see also Section 3.8.4). In this respect, of course, polymers differ significantly from metals, at least in terms of the nature of the sensitizing medium.

In general, it would seem best to assume the possibility that, unless proven otherwise, deleterious chemical effects may exist in the fatigue of polymers exposed to a given environment. Unfortunately, except for the few studies cited here and in Section 3.8.4, the possibility of environmental interaction in the fatigue of polymers has been largely neglected.

1.5 DEFORMATION AND FRACTURE MODES

1.5.1 Homogeneous Deformation and Fracture

If the stress concentration at the tip of a crack or flaw is sufficiently high, the yield stress of the material may be reached (provided that fracture does not intervene). In yielding, glassy polymers exhibit two distinct modes [25, Chap. 6; 66]: shear (or shear-resolvable) yielding, and normal-stress yielding (see Fig. 1.4). Regardless of the mode, the molecular basis is considered to be the occurrence of large-scale changes in chain conformation resulting from the cooperative motion (by rotation of C—C bonds) of chain segments. These changes may take place in a homogeneous manner throughout the body of the specimen (homogeneous shear yielding) or in an inhomogeneous manner (shear-band formation or crazing due to normal-stress yielding) (see Figs. 1.5 and 1.6).

Shear bands and crazes are similar in that a high degree of molecular orientation is developed in each. In shear bands, which usually develop at an angle between about 38 and 45° with respect to the stress axis [67–72], molecules are highly oriented at an angle between that of the band itself and the stress axis (Fig. 1.6). In crazes, on the other hand, the molecules become oriented in fibrils at approximately 0° with respect to the stress axis (Fig. 1.7). However, the two types of entity differ in that crazing involves cavitation as well as molecular orientation. Both crazes and shear bands play important roles in fracture. For example, cracks often pass through

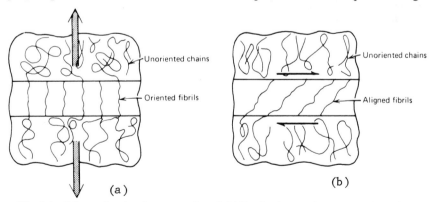

(a)   (b)

Fig. 1.4 Schema showing the two modes of yielding in glassy polymers: (a) Normal stress yielding ($|\sigma_1 - \sigma_2| = A + B/I_1$); (b) shear (or shear resolvable) yielding ($\tau_{oct} = \tau_o - \mu\sigma_m$). Note that normal stress yielding constitutes an inhomogeneous deformation that results in dilatation and crazing; shear yielding may be either a homogeneous or a heterogeneous deformation, the latter resulting in shear bands. [R. W. Hertzberg, "Deformation and Fracture Mechanics of Engineering Materials," Copyright © 1976 by John Wiley & Sons, Inc. Reprinted by permission of John Wiley & Sons, Inc.]

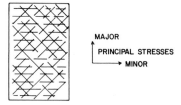

Fig. 1.5   Schema of the relative orientation of crazes (—) and shear bands (×) with respect to the principal stress axes. [After Sternstein (66).]

crazes or along shear bands [71]; crazes and shear bands may also interact, the bands serving to terminate growing crazes. (For discussions of the initiation and growth of these interesting morphological features see references [69, 70] for shear bands; [73, 74] for crazes; reference [25, Chap. 7]; and Section 2.2.4.) Sternstein [66] has argued plausibly that factors favoring uniform sequential mobility should also favor homogeneous shear yielding. Such factors are the existence of low-to-moderate stress concentrations associated with a spectrum of weak and fairly uniform flaws, low strain rates, high temperatures, and the presence of plasticizers (the last three factors tending to promote relaxation and uniform segmental mobility). In contrast, high values of stress concentrations associated with a spectrum including sharp and less uniform flaws, high strain rates, and factors hinder-

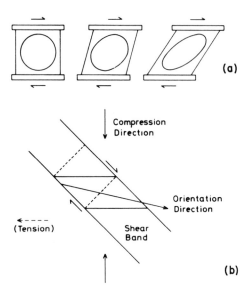

Fig. 1.6   (a) Schematic representation of simple shear deformation to large strains. (b) Relationship between compression (or tension) directions, direction of molecular orientation, and angle of shear band in localized shear yielding of a glassy polymer. [After Bucknall (25, p. 143).]

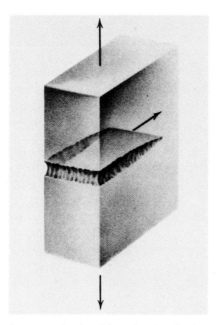

Fig. 1.7 Schema showing a craze developed from the edge of a glassy polymer subjected to a uniaxial stress in the vertical direction. Note the occurrence of both cavitation and molecular orientation, the latter in the stress direction. [R. W. Hertzberg, "Deformation and Fracture Mechanics of Engineering Materials," Copyright © 1976 by John Wiley & Sons, Inc. Reprinted by permission of John Wiley & Sons, Inc.]

ing segmental mobility, may be expected to favor inhomogeneous shear yielding in the form of shear bands (Figs. 1.4 and 1.5).

As mentioned, if normal-stress yielding is possible, a strong dilatational component results in cavitation, which, combined with localized fibrillation and orientation around the cavities, results in the formation of a craze (Fig. 1.7) at right angles to the principal stress direction (for recent detailed discussions see Andrews [15], Argon [73, 74], Kambour [75], Bucknall [25, Chapter 6]; see also Andrews [15] and Kramer [76]). As will be evident in subsequent chapters (see especially Sections 4.2.2 and 5.1.4), crazing ahead of fatigue cracks subjected to tensile loading is very common.

Whatever the detailed yielding response may be, even the most naive consideration leads to the conclusion that attainment of the critical energy (or stress-intensity factor) for crack propagation within the time scale of a deformation cycle must depend on the balance between energy stored and energy dissipated, and hence on the relative importance of the various modes of deformation and their consequences in terms of strength of the molecular aggregations that oppose the crack's progress. Clearly many possible combinations of deformations and effects may exist, thus making mechanistic correlations difficult. Thus, while fracture mechanics may serve to characterize, for example, fatigue crack propagation in terms of rates (Chapter 3), a given level of behavior may arise from a multiplicity of different competing molecular events. It becomes the task of the chemist to deduce mechanisms that are plausible on the basis of other knowledge. In this respect, it is often advantageous to adopt the viewpoint of the rheologist (who addresses the whole question of flow and its mechanisms) [66].

While our detailed knowledge of the molecular aspects of large deformations (whether in static or fatigue loading) is far from complete, it is useful to review briefly what is known (or reasonably conjectured) about the topic. For further details see Kambour and Robertson [31], Ward [32, 34], Bowden [77], Bucknall [25, Chap. 6] and, for a more recent review, Sternstein [66].

Let us now consider the case of a specimen subjected to a simple one-time tensile stress. As shown in Fig. 1.8, the response may be quite complex.

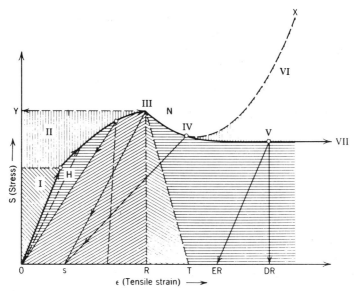

Fig. 1.8   General deformation for a polymer: stress–strain curve as observed in one material as a function of temperature or rate of strain, or for a wide variety of materials. (I) Hookean elasticity; also linear viscoelasticity; (II) nonlinear viscoelasticity; (III) yield point; (Y) yield stress; (H) hysteresis of return curves; (N) necking region; (IV) beginning of strain hardening; (S) set, permanent deformation, incipient failure; (R) rupture; (T) tearing; (V) plastic flow; (ER) partial elastic recovery, elasticoviscous, or plasticoelastic; (VI) strain hardening and rupture (VII) ductile failure. The areas I and II represent Ranier's R (to the left of the main curve); the areas below the curve minus I and II represent D. (see also Fig. 1.9). [F. R. Eirich, *Appl. Polym. Symp.* **1,** 271 (1965), Copyright © 1965 by John Wiley & Sons, Inc. Reprinted by permission of John Wiley & Sons, Inc.]

At first (region I) bonds are stretched and energy is stored in a quite reversible manner (Hookean elasticity, linear viscoelasticity). As the stress rises further, the response (region II) becomes nonlinear; energy is still recoverable on release of the stress, but slowly (anelastic response). Eventually, yielding occurs (region III), accompanied by the onset of components of irreversible deformation. (This is not to say that no reversible component exists in this region; up to some point, even crazes may exhibit a high degree of reversibility [25, p. 240; 66].) During such deformation, extensive cooperative sequential motions take place, and energy is, of course, dissipated in a viscous manner. The nominal stress, then, usually drops somewhat (region IV) and extension frequently proceeds easily, often with the formation of a neck that continuously undergoes drawing. Fracture (region V) may occur shortly after necking begins or, especially with crystalline polymers such as polyethylene, fracture may be delayed due to an orientation of the molecules (analogous to strain hardening in metals) and the stress may actually increase, as shown in region VI.

Not surprisingly, the stress–strain response depends strongly on temperature and strain rate; an increase in the former enhancing, and an increase in the latter inhibiting segmental motion [32, 33, 78]. Interestingly, the Williams–Landel–Ferry (WLF) principle of time–temperature equivalence [48, 79], which is commonly applied to characterize behavior in the range of linear response, can be applied to at least some cases involving large deformations, e.g., yielding [80, 81] and rupture [82]. Although a similar application might be valid in fatigue, e.g., to the deformation at the crack tip, insufficient data exist for analysis; moreover, the role of local hysteretic heating undoubtedly constitutes a severe complication.

In any case, under the influence of an applied stress, the energy barriers to segmental mobility are presumably decreased (cf. Eyring's transition-state model for creep [83]); the higher energy positions associated with segments that have moved in response to the stress may also correspond to higher fluidity [84]. In heterogeneous polymers, such as crystalline two-component systems, additional deformations must be kept in mind. For example, if crystallinity is present in the form of spherulites or other morphological entities, a variety of deformations within the crystallite is possible [31, 85, 86]. These include unraveling of the ordered chains, twinning, and glide along the molecular axis coupled with crystal rotation (for detailed discussions of static and fatigue behavior, see Schultz [85], Samuels [86], Friedrich [86a] and Teh and White [87]). Obviously, such deformations (Fig. 1.9) can dissipate considerable energy and the material can still remain strong if new structures (see above) are formed; in fact, crystalline polymers are among the most fatigue-resistant polymers. Whatever the detailed mechanism of deformation, the balance between energy dissipation and

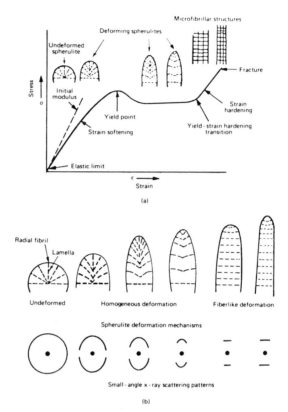

Fig. 1.9 Schemata showing deformation property changes in semicrystalline spherulitic polymers (a) morphology–strain correlation at different stages of drawing (see also Fig. 1.8); (b) morphology–small-angle x-ray patterns during drawing. [After R. J. Samuels, "Structured Polymer Properties," Copyright © 1974 by John Wiley & Sons, Inc. Reprinted by permission of John Wiley & Sons, Inc.]

strength is apparently crucial to the determination of rupture behavior. (The competitive nature of effects is discussed in Section 1.5.6.)

While different criteria have been proposed for shear yielding [77], the following criterion (a modification of the von Mises criterion) proposed by Sternstein and Ongchin [66, 88, 89] appears to be satisfactory (see Fig. 1.10):

$$\tau_{oct} \geq \tau_0 - \mu\sigma_m,\tag{1.1}$$

where τ_{oct} is the octahedral yield stress* at the yield point, τ_0 a rate- and

---

* The octahedral yield stress is the stress acting on a plane having a normal that makes the same angle to all those principal stress directions, that is, the normal defined by the equation $|\sigma_{11}| = |\sigma_{22}| = |\sigma_{33}|$ (see Fig. 1.10). The mean normal stress σ_m is given by $\sigma_m = \frac{1}{3}(\sigma_{11} + \sigma_{22} + \sigma_{33})$; σ_m should be distinguished from the mean *applied* stress that is involved in cyclic loading.

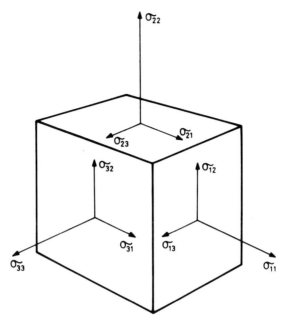

Fig. 1.10 The nine components of the stress tensor, the principal stresses being σ_{11}, σ_{22}, and σ_{33}. [From Bucknall (25, p. 151).]

temperature-dependent material constant, and μ a parameter that characterizes the sensitivity of the shear-yielding process to the mean normal stress (or hydrostatic stress) σ_m.

1.5.2 Inhomogeneous Deformation: Crazing

Hundreds of years ago, pottery glazes were observed to develop intricate, and often beautiful, patterns of fine cracks, or *crazes*.* Analysis of crazes in inorganic glasses revealed that they develop at surfaces in response to tensile stresses, usually at right angles to the principal tensile stress. Beginning nearly 40 yr ago, crazes were also noticed in polymers [91], and somewhat later Maxwell and Rahm [92, 93] and Hsiao, Marin, and Sauer [94–96] studied crazing as a function of time, temperature, stress, and environment. Even then, the need to consider crazing in design against fracture was recognized. It was also observed that crazes in polymers were not in fact simple cracks as in ceramics, and that molecular orientation and, hence,

---

* The term *crazing*, like *fracture*, has an interesting derivation from the Middle English crasen, meaning to shatter or render insane, which, in turn, is probably derived from the Old Norse krasa, meaning *to shatter* [90].

plastic deformation must be involved [94]. Since then, a great deal of work on glassy polymers, especially by Kambour [75, 97], Hull *et al.* [98–100] and Kramer [69, 76], has been devoted to elucidation of the structure within crazes, and the formation and behavior of crazes as more or less integral units. Excellent reviews by Kambour and Robertson [31], Rabinowitz and Beardmore [101], and Bucknall [25] have appeared recently; the review on deformation by Sternstein [66] also contains useful information.

Although it is not known why some polymers develop crazes more easily than others, several aspects of craze development are now well known. The development of craze surfaces at right angles to the principal tensile stress direction has been abundantly confirmed [101]; presumably less common shear-induced crazes along deformation bands have also been observed [102]. As mentioned above, development of a craze is clearly linked to the dilatational (hydrostatic) component of the applied stress [75, 101], a relationship originally put forward for cavitation in other materials by Bridgman [103]. At the same time, at least with uniaxial loads, deviatoric* stresses play a role as well [66]. Considerable evidence [66] suggests that the normal stress yielding process obeys a criterion of the form

$$|\sigma_1 - \sigma_2| \geq A + B/I_1, \tag{1.2}$$

where σ_1 and σ_2 are the major and minor principal stresses, A and B the temperature- and environment-dependent material constants, and I_1 the first stress invariant ($= 3\sigma_m$). I_1 must be positive in order to reflect the need for a dilatational (and hence cavitational) driving force. Once initiated, crazes can, of course, grow under continued stress; various expressions proposed to characterize craze growth kinetics are reviewed by Bucknall [25, Sect. 7.4].

As mentioned, the dilatational aspect of crazing has been emphasized by Gent [58], who proposed a theory for crazing in terms of a stress-induced devitrification (change from a glassy to a weaker rubbery state) at a flaw, followed by cavitation due to the hydrostatic component of the stress. He has proposed the following equation for the critical stress σ_c required to develop (and permit growth of) a thin band of softened material (which then can undergo the cavitation with deformation characteristic of crazing):

$$\bar{\sigma}_c = [\beta(T_g - T) + \sigma_m]/k, \tag{1.3}$$

where β is a coefficient relating the effect of σ_m on the glass transition temperature, T_g, T the ambient temperature, and k the stress concentration factor. Thus, crazing should be favored by high temperatures and high stress

---

* The deviatoric stress is the nonhydrostatic stress tensor that alters the shape of a specimen without altering the volume.

concentrations, and restricted by applied hydrostatic pressure. While this theory has been criticized because of the neglect of the nondilatational components of stress [66], predictions for a number of cases, including stress cracking, seem to be generally in agreement with experience.

In any case, crazing must play a major role in determining the strength of a polymer under either static or fatigue loading. Of course, an effect of frequency is to be expected, for the longer the time of load application, the greater the propensity for crazing [94]; temperatures must also be important in determining the balance between crazing and shear response [101]. In general, crazed material may be expected to be weaker than uncrazed material, though crazes are able to bear load; crazing may well constitute the ultimate limitation to static strength in a homogeneous glassy polymer. With heterogeneous systems, e.g., polymer blends, the situation is somewhat different. Here, additions of rubbery spheres in a brittle matrix are well known to exert a beneficial effect on the matrix by inducing localized ductility, in the form of crazing, around the particles (see [75] and Section 5.1.2). Rigid inclusions may also induce crazing by acting as stress concentrators [104], though this case has been little studied.

Finally, with respect to fatigue fracture, the following implications of crazing may be inferred:

(1) Crazing presumably occurs at least under some conditions in virtually all glassy polymers [31, 101] and in at least some crystalline ones [102, 105].

(2) In the static fracture of a crazable polymer, crazing is generally, if not always, a precursor of crack propagation [31, 101]. Thus fracture proceeds through crazes, and not, in general, through bulk material.

(3) While crazing provides an energy-dissipating mechanism, the strength of a specimen is lowered.

(4) Crazing may be presumed to depend on typical parameters of fatigue experiments, such as molecular weight, temperature, frequency, and environment.

For further discussion of crazing in fatigue see Chapters 4 and 5.

1.5.3 Other Effects of Cyclic Stressing

In addition to the mechanical effects described above, two other effects that involve a change in state are sometimes observable in fatigue experiments, namely, a change in modulus or in temperature. If these effects occur to any significant extent, they will alter the mechanical response at the crack tip, whether the response be homogeneous (e.g., shear yielding) or heterogeneous (e.g., crazing).

If fatigue loading involves relatively low loads (relative to the fracture stress) except at the immediate region of a flaw or crack tip, one would not expect to find changes in the properties of the bulk material. Thus in the fatigue of several polymers at low stresses and strains, little or no effect on viscoelastic response has been noted by Kodama [106] and Murukami *et al.* [107]. At the same time, changes have been reported in several cases when net section plastic strains are introduced. For example, Beardmore and Rabinowitz [23] have noted that softening (i.e., a lowering of the yield strength) under cyclic loading at a constant strain range is characteristic of unnotched glassy linear polymers (see Section 2.5); this effect appears to be associated with crazing (see Section 5.1.4). In contrast, evidence for stiffening also exists. For example, increases in modulus have been observed [108–110] in several polymers (perhaps due to changes in crosslinking). More recently, Bouda *et al.* [111] described slight, progressive increases in Young's modulus when PMMA and nylon 6 were fatigued in torsion (at 1 Hz and at a constant stress range) for up to 10^7 cycles, accompanied by decreases in internal damping. Interestingly, these changes took place before crazing occurred. At the same time, densities *decreased* significantly on cycling, an observation *apparently* inconsistent with the increase in modulus noted. To explain this anomaly, Bouda *et al.* [111] proposed that while the *average specific volume* is *increased* by fatigue (i.e., yielding a lower density, a change that is intuitively reasonable), *local volume contraction* may occur. (Differences of this kind may well exist, for volume and enthalpy may relax at different rates [112].) Thus it seems likely that some kind of mechanical conditioning is involved in such cases; for additional evidence of conditioning during repetitive strain cycles using PMMA, see Ho [113] and Sternstein [66].

The state of a polymer may also be changed by imposing even a subtly different thermal history [112]. While the effects on fatigue are not yet elucidated due to thermomechanical history, the general phenomenon of aging must be taken into account in any consideration of deformation and fracture. With respect to aging, Struik [112] has put forward an interesting speculation to account for the fact that creep is greater with intermittent than with a static load of the same magnitude (see also Section 5.1.5). He argues that the increase in compliance in monotonic creep is opposed by a small effect of aging with associated relaxation of volume enthalpy, and that the repeated stressing in intermittent creep erases at least part of the aging effect that would otherwise oppose the decrease in compliance. (Note that this effect is the reverse of that discussed by, e.g., Sternstein [66].) In any case, contrary effects of mechanical history can apparently exist.

As mentioned above, the state of a polymer can also be perturbed by adiabatic heating (see Section 2.3.2 for a detailed discussion). With lossy polymers (high tan δ), pseudoadiabatic heating can arise due to a combination of hysteretic heating with low conduction of heat away from the specimen. Such heating can not only change the state of the polymer, but also result in gross weakening, and even melting [22]. In fact, fatigue failure of this type has been very much emphasized in theory and in testing practice (see Section 2.3.2). At the same time, as will be shown later, heating localized at the crack tip may induce relaxation and thus reduce the concentration of stress. Apart from this cyclic hysteretic heating, heating could also arise, in principle, from the release of strain energy as fresh crack surfaces are formed. However, Kambour [114] has shown that a negligible effect will be observed as long as the growth rate of a crack is less than 1 cm/sec in a typical glassy polymer. Thus, when fatigue cracks are growing in a stable manner, this contribution can probably be neglected.

More subtle effects have been noted in polycarbonate by Higuchi [115], who observed an initial *cooling* during the first few cycles in fatigue tests. Higuchi attributed the cooling to a thermoelastic effect; he also developed relationships between the temperature rise and anelastic strain.

1.5.4 Factors Associated with Toughness and Ductility

When stressed, polymers exhibit behavior ranging from a ductile to a brittle (yielding and flow versus facile crack initiation and propagation, respectively) response. The external factors favoring yielding have already been discussed: a high temperature and a low strain rate. As shown in Table 1.1, molecular structure is also important in determining the relative tendencies toward ductility or brittleness. While Table 1.1 is based on experiments involving monotonic loads, it might be expected that resistance to fatigue should parallel resistance to static fracture. Indeed, this is often so (see Section 3.8.7).

Since no analytical theory exists that will permit the quantitative prediction of ductility, one must seek correlations between ductility and other properties, and then try to relate these properties to structure. Although no correlation should be expected to hold in all cases, several generalizations do appear to be useful. Two of the more plausible correlations, with secondary transitions and with free volume, are briefly discussed below; their application to the case of fatigue fracture has received little attention. As will become obvious, the correlations are often themselves related.

One of the most studied (and controversial) correlations of ductility with other properties has been based on viscoelastic relaxation behavior, and especially on the nature of relaxations characterized by secondary transitions in complex modulus (or compliance). Thus about 15 yr ago, while studying

TABLE 1.1   Ductile–Brittle Behavior and the Stress Configuration of
Some Commercial Plastics[a]

Stress configuration	$(\sigma_1 - \sigma_3)/2\sigma_1$	Polymers ductile at 25°C and 1 min^{-1} strain rate	
		Glassy polymers	Crystalline polymers
Stress around Izod notch	$<\frac{1}{2}$	Cellulose nitrate	Low-density polyethylene
		Cellulose propionate	Ethylene-propylene
		Bisphenol-A polycarbonate	copolymers
		Vinyl chloride-vinyl acetate copolymers	
Tension	$\frac{1}{2}$	All of the above plus	All of the above plus
		Poly (2,6-dimethyl-1, 4-phenylene oxide)	Acetal
			Nylon
		Poly(ethylene terephthalate)	High-density polyethylene
		Polysulfone, aryl	Poly(4-methyl pentene-1)
			Polypropylene
Simple shear	1	All of the above plus	All of the above plus
		Poly(methyl methacrylate)	Poly(ethylene terephthalate)
Compression	(∞)	All of the above plus	All of the above
		Polystyrene	
		Phenolics	

[a] From Kambour and Robertson [31], with permission from North-Holland Publ.

the damping behavior of polymers, Oberst [116] noted a correlation between a transition in impact strength (from low to high values) and a temperature corresponding closely to the temperature of a major damping peak (below room temperature), at a frequency of about 10 Hz. Similar observations have since been made for many polymers, and for a variety of large-deformation processes, including yielding [81] and failure itself [80, 117–120]. Much evidence has been reviewed by Boyer [117], and Heijboer [120] has conducted systematic studies on the effects of molecular structure.

For example, as shown by Retting [80] and Bauwens [81], the yield stress of poly(vinyl chloride) is related not only to the position of the secondary transition (the so-called β-transition) but also to its intensity. Broutman and Kobayashi [121] interpreted the fracture energy of a growing crack in poly(methyl methacrylate) in terms of motions associated with transition temperatures. Johnson and Radon have reported [118, 119] that the time-to-failure in poly(methyl methacrylate) and poly(vinyl chloride) (at the ductile–brittle transition temperature) corresponded approximately to the relaxation time (from mechanical or dielectric spectra) associated with the β-process at the same temperature.

In any case, correlations of the kind discussed are not *necessarily* characteristic of all polymers, for two reasons. First, in contrast to earlier ideas, it is clear that the mere existence of a significant low-temperature transition is, while perhaps necessary for ductility, not necessarily sufficient. Thus, in his detailed studies of a series of substituted acrylic polymers, Heijboer [120] showed an effective secondary transition must involve main chain segmental motion rather than the mere rotation of side groups. Certainly the latter motion could play only a second-order role in the gross deformation of a polymer. Second, as pointed out by Vincent [122] and Heijboer [120], there is no a priori reason why a low-strain, linear phenomenon such as damping must be directly related to an ultimate property such as fracture. Nevertheless such a correlation has apparently been identified in some cases.

At the same time, Knauss [123] has observed that the relationships of linear viscoelasticity may, in fact, apply to a significant extent to cases of nonlinear deformation. Indeed the applicability of the WLF time–temperature relationship to failure as well as to linear viscoelastic processes was mentioned above [31]. The reason is, undoubtedly, that it is the segmental motions that are important in each case; the controlling molecular process at a crack tip may well be the same process that is operative in the small-deformation measurement of damping. The fact that a damping peak exists at a low temperature means that its characteristic motion (e.g., of a short-chain segment) is relatively free above that temperature.

The fact that the modulus decreases as the temperature is raised through a transition was emphasized long ago by Bobalek [124], who showed a correlation between the relative modulus *decrease* (over a temperature range from below to slightly above ambient temperature) and an *increase* in impact strength. A similar point was made later by Kambour and Robertson [31]: if the yield stress decreases in parallel with the modulus as a function of temperature, while the crazing stress remains relatively constant, ductility should be favored at higher temperatures.

In the case of fatigue, some correlation between crack growth behavior and viscoelastic character is also possible. For example, as discussed in detail in Section 3.4, the sensitivity of fatigue crack growth in polymer to frequency is clearly related to the relationship between the temperature and frequency of the test itself and those of the β-transition. However, beyond this generalization, the evidence is fragmentary and often contradictory. Conflicting results have been found by several investigators in studies of fatigue crack growth in several polymers [125, 126 and Section 3.5]; crack growth rates as a function of temperature paralleled tan δ curves in some cases but not in others. Thus the relationship between viscoelastic response and fatigue behavior is not yet established.

Whatever the connection between viscoelastic characteristics and failure, one would expect free volume to play a complementary role. This expectation was confirmed by Litt and Tobolsky [127] who obtained an interesting correlation between impact strength and the "maximum fractional unoccupied volume" $\bar{f}$:

$$\bar{f} = 1 - (d_a/d_c), \tag{1.4}$$

where d_c and d_a are the densities of the crystalline and the amorphous phases, respectively (d_c corresponding to the theoretical crystalline density). When $\bar{f}$ is high, say 0.09 or more, ductility tends to be high, as in polycarbonate; when $\bar{f}$ is less than about 0.09, ductility tends to be low, as in polystyrene. The basis for this correlation (which, unfortunately, is not universally valid) probably lies in the increased difficulty of flow as free volume is reduced [31]. If, then, free volume is reduced for any reason, e.g., by annealing or the introduction of so-called antiplasticizers [31, 63], the yield stress and the probability of fracture will likely be increased while toughness is reduced.

A related approach has been based on measurements of the coefficient of expansion. Kambour and Robertson [31] have suggested that a correlation may exist between ductility and the factor $(a_r - a_g)$, the difference between the coefficients of expansion in the glassy and rubbery states, and a measure of excess free volume. The crucial role of preexistent mobile free volume has also been emphasized in the detailed review of strength behavior by Eirich [128]; the need for dilatational effects to create the free volume needed for segmental flow is presumably reduced in such a case.

Returning to the role of transitions, it seems likely that the common association of low-temperature transitions with ductility may simply reflect the fact that at temperatures above, for example the β transition, free volume increases to the point that segmental mobility and relaxation become possible (see also the discussion by Struik [112]).

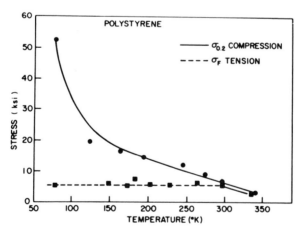

Fig. 1.11 Effect of temperature on the 0.2% flow stress (———) and the fracture stress (– – –) of PS. [Reprinted with permission from S. Rabinowitz and P. Beardmore, *CRC Crit. Rev. Macromol. Sci.* **1**, 1 (1972). Copyright © The Chemical Rubber Co., CRC Press, Inc.]

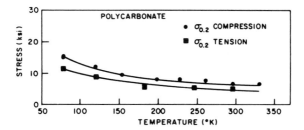

Fig. 1.12   Effect of temperature on the 0.2% flow stress of PC in tension (■) and compression (●). [Reprinted with permission from S. Rabinowitz and P. Beardmore, *CRC Crit. Rev. Macromol. Sci.* **1,** 1 (1972). Copyright © The Chemical Rubber Co., CRC Press, Inc.]

1.5.5   Crazing versus Shear

Since crazing is a weakening process competitive with shear yielding [25, Chapter 6; 31], it is important to consider the effects of temperature on the balance between the two responses. Rabinowitz and Beardmore [100] have shown that in the temperature region of semiductility, both tensile and compressive flow (or yield) stresses have similar temperature dependences (e.g., at temperatures of 300°K and above in polystyrene (Fig. 1.11); however, at low temperatures, the fracture stress is lower and the material fails before yielding. If crazing plays a dominant role in fracture, then any factor that would raise the critical craze stress should raise the lower curve in Fig. 1.11 (i.e., the fracture stress curve) closer to the yield curves. Reduction in the yield stress' dependence on temperature would have a similar effect. Indeed, polycarbonate exhibits such a reduced slope of yield stress versus temperature; the yield stress approaches the crazing stress over a wide range of temperature (Fig. 1.12). Thus, the relative magnitudes of flow (yield) and crazing stresses are undoubtedly responsible for some of the effects of temperature on ductility, and may well be important in fatigue response.

1.5.6   Competitive Events in Fracture

Whatever the precise mechanisms that control ductility, significant effects of structure and morphology exist. Ductility tends to decrease as one moves upward in Table 1.1, i.e., as the stress field becomes increasingly severe. In general, maximum ductility in amorphous polymers tends to occur in molecules possessing bulky repeating units *in the main chain*, e.g., phenyl groups or cellulosic rings. This pattern is reasonably consistent with the notions of free volume and secondary relaxations previously discussed. In the case of crystalline polymers, the possibilities of lamellar and spherulitic deformations add in, effect, additional modes of ductile response (shear

yielding) so that the onset of fracture can be delayed to large strains (see
Fig. 1.8 and the discussion below). Crazing may occur in some crystalline
polymers as well [102, 106]. All these processes that contribute to ductility
(in the sense of a significant elongation prior to fracture) compete with the
process of fracture; indeed, even an apparently brittle fracture clearly involves a surprisingly high degree of plastic deformation.

Vincent [78] and Eirich [77b] provide excellent discussions of the nature
of the competition between fracture and deformation without fracture. For
example, the dilatational component of a tensile stress enhances relaxation,
but also the opening (cavitation) or growth of flaws. Cold-drawing relieves
stresses, but may lead to a weak structure, and so on. Semicrystalline polymers (already given as an example of typically ductile materials) provide an
interesting example of the interplay between opposing tendencies.

In Fig. 1.13, Meinel and Peterlin [128] have visualized the stress–strain
curve (of a semicrystalline polymer) as the resultant of two other curves,
one corresponding to the breakdown of the original spherulitic structure,

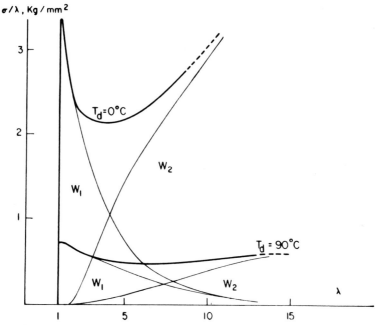

Fig. 1.13 Separation of the differential work density σ/λ into the two components W_1 and
W_2. λ represents the draw ratio, W_1 represents the work of applied forces during elastic deformation, lamellar slipping and fracture, and W_2 represents the deformational work of the fibrous
structure formed on drawing. T_d represents the temperature of draw. [G. Meinel and A. Peterlin,
J. Polym. Sci. part A-2, **9,** 67 (1971), Copyright © 1971 by John Wiley & Sons, Inc. Reprinted by
permission of John Wiley & Sons, Inc.]

and the other corresponding to the organization of the broken down material into new, oriented, and strong units. If after yielding, the stress drops as a function of elongation, the rheological state becomes unstable, and the material becomes more prone to fracture [78]; the greater the drop in stress during drawing, the greater the instability. For the attainment of maximum strength, this weakening must be successfully opposed by structural reorganization (e.g., by strain hardening); such reorganization must be possible on the time scale of the experiment before fracture takes place. An analogous competition between fracture and the buildup of a new structure also may be presumed to exist in amorphous polymers.

Evidently this is often the case with crystalline polymers, whose high fatigue resistance in general has been noted above. (The role of crystallinity is not confined to plastics; Andrews [129] has shown that fatigue resistance is enhanced by the ability of a rubber to crystallize at the crack tip.)

In assessing the likelihood of fracture in polymers having various structures, compositions, and morphologies, the three simple operational principles enunciated by Vincent [78] can serve as a useful guide:

(1) the higher the stress at a given elongation, the greater the probability of fracture in a specimen;

(2) the higher the concentration of stress, the greater the probability of fracture; and

(3) the smaller the drop in stress on yielding, the greater the resistance to fracture, and vice versa.

Of course, it is implicit that energy dissipated is energy unavailable for release by the formation of new crack surface, assuming that hysteretic heating does not lower hydrodynamic stability.

Several typical applications of these principles may be given. For example, necking may *reduce* fracture resistance if the segmental flow becomes unstable (as shown by a drop in stress as a function of elongation). A lowering of molecular weight reduces strength by increasing stress concentrations and by inhibiting molecular orientation that would increase strength. Conversely, an increase in molecular weight should permit the establishment of a more stable entanglement network that could enhance strength (see Section 3.8.2). In all such considerations, it must be kept in mind that a given parameter may have contradictory effects with respect to the three principles. For a detailed discussion of specific effects of structure and composition on strength in terms of the three principles, see Vincent [78].

In fatigue, an additional parameter must be considered, the effect of the time scale within a cycle (both rise time and duration). Thus frequency may have several effects, e.g., an increase in modulus, change in state, and a change in the balance between shear yielding and crazing depending on the polymer.

In practice, increasing the frequency may result in no change, an increase, or a decrease in fatigue crack growth rates (see Section 3.4). Wave form will also be important, for changing the rise time is equivalent to changing the strain rate, thus altering the balance between competitive processes.

In the following chapters, examples of many of the topics discussed above will be discussed in more detail: in particular, fatigue testing, the effects of molecular properties and composition, the micromechanisms of fatigue fracture, and the complex-behaving composites.

REFERENCES

[1] "Oxford English Dictionary," Vol. IV, p. 101. Oxford Univ. Press (Clarendon), London and New York, 1933.

[2] M. D. Skibo, J. Janiszewski, R. W. Hertzberg, and J. A. Manson, *Proc. Int. Conf. Toughening Plast.*, *London* Paper 25 (July 1978).

[3] P. M. Clarke, *Plast. Rubber Inst. Conf.*, *Cranfield, England, January 8–10* (1973), quoted by G. B. Bucknall and W. W. Stevens, *in* "Toughening of Plastics," Paper 24, Plastics and Rubber Institute, London, 1978.

[4] A. Wöhler, *Engl. Abstr. Eng.* **2**, 199 (1871).

[5] R. W. Hertzberg, "Deformation and Fracture Mechanics of Engineering Materials." Wiley, New York, 1976.

[6] R. P. Kambour, *Am. Chem. Soc. Polym. Preprints* **12**, 52 (1971).

[7] E. H. Andrews, *J. Mat. Sci.* **9**, 887 (1974).

[8] J. G. Williams, *Adv. Polym. Sci.* **27**, 67 (1978).

[8a] J. H. Dillon, Fatigue phenomena in high polymers, *Adv. Colloid Sci.* **3**, 219 (1950).

[9] B. Rosen, "Fracture Processes in Polymeric Solids." Wiley (Interscience), New York, 1964.

[10] I. Wolock and S. Newman in "Fracture Processes in Polymeric Solids" (B. Rosen, ed.) p. 235. Wiley (Interscience), New York, 1964.

[11] R. F. Landel and R. F. Fedors, "Fracture Processes in Polymeric Solids" (B. Rosen, ed.), p. 361. Wiley (Interscience), New York, 1964.

[12] E. H. Andrews, *in* "Physical Basis of Yield and Fracture," Conf. Ser. No. 1, p. 127. Institute of Physics and Physics Society, London, 1966.

[13] E. H. Andrews, "Fracture of Polymers." American Elsevier, New York, 1968.

[14] E. H. Andrews, *in* "Testing of Polymers" (W. Brown, ed.), Vol. IV, p. 237. Wiley (Interscience), New York, 1969.

[15] E. H. Andrews, *in* "The Physics of Glassy Polymers" (R. N. Haward, ed), p. 394. Wiley, New York, 1973.

[16] C. B. Bucknall, K. V. Gotham, and P. I. Vincent, *in* "Polymer Science. A Materials Handbook" (A. D. Jenkins, ed.), Vol. I, Chapter 10. American Elsevier, New York, 1972.

[17] P. I. Vincent, *Encycl. Polym. Sci. Technol.* **7**, 292 (1968).

[18] J. W. S. Hearle, *J. Mater. Sci.* **2**, 474 (1967).

[19] V. R. Regel and A. M. Leksovskii, *Mekh. Polim.* **5**, 70 (1965).

[20] V. P. Tamuzh, *Mekh. Polim.* **5**, 97 (1965).

[21] J. H. Plumbridge, *J. Mater. Sci.* **7**, 939 (1972).

[22] J. A. Manson and R. W. Hertzberg, *Crit. Rev. Macromol. Sci.* **1**, 433 (1973).

[23] P. Beardmore and S. Rabinowitz, *in* "Plastic Deformation of Materials" (R. J. Arsenault, ed.), Treatise on Materials Science and Technology, p. 267. Academic Press, New York, 1975.

[24] J. M. Schultz, *in* "Properties of Solid Polymeric Materials" (J. M. Schultz, ed.), Part B, Treatise on Materials Science and Technology, p. 599. Academic Press, New York, 1977.

[25] C. B. Bucknall, "Toughened Plastics." Applied Science Publ., London, 1977.

[25a] C. B. Bucknall, *Adv. Polym. Sci.* **27**, 121 (1978).

[25b] J. A. Manson and L. H. Sperling, "Polymer Blends and Composites." Plenum Press, New York, 1976.

[26] M. J. Owen, Fracture and Fatigue, *in* "Composite Materials," Vol. 5, Chapters 7 and 8. Academic Press, New York, 1974.

[27] B. Harris, *Composites* **8**, 214 (1977).

[28] E. H. Andrews and P. E. Reed, *Adv. Polym. Sci.* **27**, 1 (1978).

[29] H. H. Kausch, "Polymer Fracture," Vol. 2, Polymers Properties and Applications. Springer-Verlag, New York, 1978.

[30] J. P. Berry, Brittle behavior of polymeric solids, *in* "Fracture Processes in Polymeric Solids" (B. Rosen, ed.), p. 195. Wiley (Interscience), New York, 1964.

[31] R. P. Kambour and R. E. Robertson, *in* "Polymer Science. A Materials Science Handbook" (A. D. Jenkins, ed.), p. 687. North Holland Publ., Amsterdam, 1972.

[32] I. M. Ward, *J. Mater. Sci.* **6**, 1397 (1971).

[33] I. M. Ward, "Mechanical Properties of Solid Polymers." Wiley (Interscience), New York, 1971.

[34] J. R. Beatty, *Rubber Chem. Technol.* **37**, 1341 (1964).

[35] G. J. Lake and P. B. Lindley, *in* "Physical Basis of Yield and Fracture," Conf. Ser. No. 1, p. 176. Institute of Physics and Physics Society, London, 1966.

[36] G. J. Lake, *Rubber Age* **8**, 30 (1972); **10**, 39 (1972).

[37] A. N. Gent, *in* "Fracture, an Advanced Treatise" (H. Liebowitz, ed.), Vol. VII, p. 316. Academic Press, New York, 1972.

[38] F. R. Eirich and T. L. Smith, *in* "Fracture, an Advanced Treatise" (H. Liebowitz, ed.), Vol. VII, p. 476. Academic Press, New York, 1972.

[39] F. W. Billmeyer, Jr., "Textbook of Polymer Science," 2nd ed. Wiley (Interscience), New York, 1971.

[40] F. Rodriguez, "Principles of Polymer Systems." McGraw-Hill, New York, 1970.

[41] L. E. Nielsen, "Mechanical Properties of Polymers and Composites," Vols. 1 and 2. Dekker, New York, 1974.

[42] Anonymous, "Modern Plastics Encyclopedia," p. 463. McGraw-Hill, New York, 1978–1979 ed.

[43] K. Oberbach, *Kunstoffe* **63**(1), 35 (1973).

[44] T. P. Oberg, R. T. Schwartz, and D. A. Shinn, *Mod. Plast.* **20**(8), 87 (1943).

[45] K. N. Lauraitis, ASTM STP, p. 267 (1977).

[46] P. C. Paris and G. C. M. Sih, STP 381, p. 30 (1965).

[47] S. J. Hutchinson and P. P. Benham, *Plast. Polym.* **38**, 102 (1970).

[48] M. L. Williams, R. F. Landel, and J. D. Ferry, *J. Am. Chem. Soc.* **77**, 3701 (1955).

[49] G. S. Y. Yeh, *Crit. Rev. Macromol. Sci.* **1**, 173 (1972).

[50] A. Casale and R. S. Porter, "Polymer Stress Reactions," Vols. 1 and 2. Academic Press, New York, 1978.

[51] S. N. Zhurkov and E. E. Tomashevskii, "Physical Basis of Yield and Fracture," Conf. Ser. 1. Inst. Phys. Physical Soc., Oxford, 1966.

[51a] M. L. Williams and K. L. DeVries, *Proc. Int. Congr. Rheol., 5th* **3**, 139 (1970).

[52] R. Brown, K. L. DeVries, and M. L. Williams, *in* "Polymer Networks: Structural and Mechanical Properties" (A. J. Chompff, ed.), p. 409. Plenum Press, New York, 1971.

[53] K. L. DeVries, B. A. Lloyd, and M. L. Williams, *J. Appl. Phys.* **42**(12), 4644 (1971).

[54] K. L. DeVries, D. K. Roylance, and M. L. Williams, *Int. J. Fracture Mech.* **7**(2), 197 (1971).

[55] B. A. Lloyd, K. L. DeVries, and M. L. Williams, *J. Polym. Sci. Polym. Lett. Ed.* **10**, 1415 (1972).

[56] A. Peterlin, *J. Polym. Sci. Polym. Lett. Ed.* **7**, 1151, (1969).

[57] G. P. Marshall and J. G. Williams, *Proc. R. Soc. London Ser. A* **342**, 55 (1975).

[57a] G. P. Marshall, L. E. Culver, and J. G. Williams, *Proc. R. Soc. London Ser. A* **319**, 165 (1970).

[58] A. N. Gent, *J. Mater. Sci.* **5**, 925 (1970).

[59] R. P. Kambour, C. L. Gruner, and E. E. Romagosa, *J. Polym. Sci. Polym. Phys. Ed.* **11**, 1879 (1973).

[60] E. H. Andrews, G. M. Levy, and J. Willis, *J. Mater. Sci.* **8**, 1000 (1973).

[61] N. Brown, *J. Polym. Sci. Polym. Phys. Ed.* **11**, 2099 (1973).

[62] Y. M. Mai and J. G. Williams, *J. Mater. Sci.* **14**, 1933 (1979).

[63] L. M. Robeson and J. A. Faucher, *J. Polym. Sci. Polym. Phys. Ed.* **7**, 35, (1969).

[64] D. G. LeGrand, *J. Appl. Polym. Sci.* **13**, 2129 (1969).

[65] P. E. Bretz, R. W. Hertzberg, J. A. Manson, and A. Ramirez, ACS Symposium Series (1980) (in press).

[66] S. S. Sternstein, *in* "Properties of Solid Polymeric Materials" (J. M. Schultz, ed.), Part B, Treatise on Materials Science and Technology, p. 541. Academic Press, New York, 1977.

[67] W. Whitney, *J. Appl. Phys.* **34**, 3633 (1963).

[68] A. S. Argon, R. D. Andrews, J. A. Godrick, and W. Whitney, *J. Appl. Phys.* **39**, 1899 (1968).

[69] E. J. Cramer, *J. Polym. Sci. Polym. Phys. Ed.* **13**, 509 (1975).

[70] P. B. Bowden and S. Raha, *Phil. Mag.* **22**, 463 (1970).

[71] L. Camwell and D. Hull, *Phil. Mag.* **27**(5), 1135 (1972).

[72] T. E. Brady and G. J. Y. Yeh, *J. Mater. Sci.* **8**, 1083 (1973).

[73] A. S. Argon and J. G. Hannoosh, *Phil. Mag.* **36**(5), 1195 (1977).

[74] A. S. Argon and M. M. Salama, *Phil. Mag.* **36**(5), 1217 (1977).

[75] R. P. Kambour, *J. Polym. Sci. Macromol. Rev.* **7**, 1 (1973).

[76] E. J. Kramer, *in* "Developments in Polymer Fracture" (E. H. Andrews, ed.). Applied Science Publ., London, 1979.

[77] P. B. Bowden, *in* "The Physics of Glassy Polymers" (R. N. Haward, ed), p. 279. Wiley, New York, 1973.

[77a] P. B. Bowden and J. A. Jukes, *J. Mater. Sci.* **3**, 183 (1968).

[77b] F. R. Eirich, *Appl. Polym. Symp.* **1**, 271 (1965).

[78] P. I. Vincent, "Physical Basis of Yield and Fracture." Inst. Phys. Physical Soc., Oxford, 1966.

[79] J. D. Ferry, "Viscoelastic Properties of Polymers," 2nd ed. Wiley, New York, 1970.

[80] W. Retting, *Eur. Polym. J.* **6**, 853 (1970).

[81] J. C. Bauwens, *J. Polym. Sci. Polym. Symp.* **33**, 123 (1971).

[82] F. Bueche and J. Halpin, *J. Appl. Phys.* **35**, 36 (1964).

[83] H. Eyring, *J. Chem. Phys.* **4**, 283 (1936).

[84] R. E. Robertson, *Appl. Polym. Symp.* **7**, 201 (1968).

[85] J. M. Schultz, "Polymer Materials Science." Prentice-Hall, Englewood Cliffs, New Jersey, 1973.

[85a] J. Magill, *in* "Properties of Solid Polymeric Materials" (J. M. Schultz, ed.), Part A, Treatise on Materials Science and Technology, p. 1. Academic Press, New York, 1977.

[86] R. S. Samuels, "Structured Polymer Properties." Wiley, New York, 1974.

[86a] K. Friedrich, *Proc. Int. Conf. Fracture, 4th, Waterloo, June* (D.M.R. Taplin, ed.), **3**, 1119 (1977).

[87] J. R. White and J. W. Teh, *Polymer* **20**, 764 (1979).

[88] S. Sternstein and L. Ongchin, *ACS Polym. Preprints* **10**, 1117 (1969).

[89] S. Sternstein, L. Ongchin, and A. Silverman, *Appl. Polym. Symp.* **7**, 175 (1969).

[90] "Oxford English Dictionary," Vol. II, p. 1147. Oxford Univ. Press (Clarendon), London and New York, 1933.

[91] W. B. Klemperer, *Appl. Mech.* 328 (1941).

[92] B. Maxwell and L. F. Rahm, *Ind. Eng. Chem.* **41**, 1988 (1949).

[93] B. Maxwell and L. F. Rahm, *J. Soc. Plast. Eng.* **6**, 7 (1950).

[94] J. A. Sauer, J. Marin, and C. C. Hsiao, *J. Appl. Phys.* **20**, 507 (1949).

[95] C. C. Hsiao and J. A. Sauer, *J. Appl. Phys.* **21**, 1071 (1950).

[96] J. A. Sauer and C. C. Hsiao, *Trans. ASME* **75**, 895 (1953).

[97] R. P. Kambour, *Polymer* **5**, 143 (1964).

[98] J. Murray and D. Hull, *Polymer* **10**, 451 (1969).

[99] M. Bevis and D. Hull, *J. Mater. Sci.* **5**, 983 (1970).

[100] D. Hull, *J. Mater. Sci.* **5**, 357 (1970).

[101] S. Rabinowitz and P. Beardmore, *CRC Crit. Rev. Macromol. Sci.* **1**, 1 (1972).

[102] J. J. Harris and I. M. Ward, *J. Mater. Sci.* **5**, 573 (1970).

[103] P. W. Bridgman, "Studies in Large Plastic Flow and Fracture with Special Emphasis on the Effects of Hydrostatic Pressure," 1st ed. McGraw-Hill, New York, 1952.

[104] R. R. Lavengood, L. Nicolais, and M. Narkis, *Tech. Rep. AD891*, p. 254. Nat. Tech. Inf. Service (November 1971).

[105] E. A. Flexman, *Proc. Int. Conf. Toughening Plast.*, *London* Paper 14 (July 1978).

[106] M. Kodama, *J. Appl. Polym. Sci.* **20**, 2853 (1976).

[107] R. Murukami, N. Shin, N. Kusomoto, and Y. Motozato, *Kobunshi Robunshu* **33**, 107 (1976).

[108] M. Schrager, *J. Polym. Sci. Polym. Lett. Ed.* **8**, 1999 (1970).

[109] J. O. Outwater and M. C. Murphy, *Proc. Ann. Tech. Conf.*, *26th, SPI*, 1971.

[110] S. Shimamura and H. Maki, *Proc. Jpn. Congr. Test. Mater.*, *5th*, **5**, 136 (1962).

[111] V. Bouda, V. Zilvar and A. J. Staverman, *J. Polym. Sci.*, *Polym. Phys. Ed.* **14**, 2313 (1976).

[112] L. C. E. Struik, "Physical Aging in Amorphous Polymers and Other Materials." American Elsevier, New York, 1978.

[113] T. C. Ho, Ph.D. dissertation, Rensselaer Polytechnic Inst., Troy, New York, 1973, quoted in Sternstein [66, pp. 562–564].

[114] R. P. Kambour and R. E. Barker, Jr., *J. Polym. Sci. Polym. Lett. Ed.* **4**, 35 (1966).

[115] M. Higuchi and Y. Imai, *J. Appl. Polym. Sci.* **14**, 2377 (1970).

[116] H. Oberst, *Kunstoffe* **53**, 1 (1963).

[117] R. F. Boyer, *Polym. Eng. Sci.* **8**, 161 (1968).

[118] F. A. Johnson and J. C. Radon, *Eng. Fracture Mech.* **4**, 555 (1972).

[119] J. C. Radon, *Polym. Eng. Sci.* **12**, 425 (1972).

[120] J. Heijboer, *J. Polym. Sci. Polym. Symp.* **16**, 3755 (1968).

[121] L. J. Broutman and T. Kobayashi, *Int. Conf. Dyn. Crack Propagat.*, *Lehigh Univ.*, *July 10–12* (1972).

[122] P. I. Vincent, *Polymer* **1**, 425 (1960).

[123] W. G. Knauss, *Int. J. Fracture Mech.* **6**, 7 (1970).

[124] E. G. Bobalek and R. M. Evans, *SPE Trans.* **1**, 93 (1961).

[125] G. C. Martin and W. W. Gerberich, *J. Mater. Sci.* **11**, 231 (1976).

[126] S. Arad, J. C. Radon and L. E. Culver, *J. Appl. Polym. Sci.* **17**, 1467 (1973).

[127] M. H. Litt and A. V. Tobolsky, *J. Macromol. Sci.-Phys.* **B1**(3), 433 (1967).

[128] G. Meinel and A. Peterlin, *J. Polym. Sci. Polym. Lett. Ed.* **9**, 67 (1971).

[129] E. H. Andrews, *J. Appl. Phys.* **32**, 542 (1961).

2 | *Cyclic Stress and Strain Fatigue: Unnotched Test Specimens*

2.1 INTRODUCTION

Polymeric solids are being utilized in the manufacture of an increasing number of load-bearing structural components. As examples, plastics are finding increased usage in water and gas transport systems, gears, hinges, springs, and mechanical arms. In part, this has resulted from an energy-related trend toward lighter-weight engineering systems and a steady improvement in the mechanical properties of many plastics. To provide assurances that these parts will withstand the rigors associated with their service life, more detailed characterizations of deformation and fracture properties of polymeric solids are demanded. Since many loads are cyclic in nature, the deformation and fracture response to cyclic loads (fatigue) is of particular interest.

Engineers and designers are well aware of the insidious nature of structural damage resulting from repetitive loadings; while one load excursion may not cause fracture, repeated stressing to the same load level (even less than the nominal yield strength) will precipitate damage and eventual failure. This phenomenon has long been known to engineers and indicates that some irreversible damage is introduced as a result of each loading excursion. Failure then occurs when some critical level of damage is accumulated to either fracture the component or render it incapable of satisfactorily perform-

34

ing its intended function. The actual product service life will then depend strongly on the number of stress cycles experienced during its intended life span and the extent of damage accumulated per loading excursion.

For the past one hundred years, engineers have sought to identify the major variables responsible for fatigue failures. A focal point to these studies has been the realization that fatigue damage will occur only when cyclic plastic* strains are generated. That fatigue fractures do occur when nominal applied stresses are below the material yield strength reflects the fact that surface or internal stress concentrations are present, which readily elevate local stresses and associated strains into the plastic range. Since preexistent defects are sometimes encountered, it is often desirable to evaluate the fatigue resistance of a component in terms of either crack initiation or subsequent crack propagation. For example, in the case of a component that possesses a simple configuration and was manufactured under carefully controlled conditions, the presence of an adventitious defect may not represent a highly probable event. Consequently, it would be reasonable to make useful service life estimates for the part in question from experimental data derived from unnotched laboratory samples. Here most of the fatigue life would be consumed during crack initiation. By contrast, when it is likely that a design, manufacturing or service-induced defect exists in the component, or that failure of the component would lead to dangerous consequences, service life estimations should be dominated by crack propagation considerations. In this instance, it would make more sense to base service life design decisions on data derived from precracked samples.

Herein lies a major design dilemma. Should one assume that a certain structural component does or does not contain defect(s) when first placed in service? If the likelihood of a starting defect is ignored, a useful service life might be construed to be the time or number of loading cycles needed to generate a crack of some small but finite dimension. In essence, this is the tack taken by most automotive design engineers. By sharp contrast, most aeronautical and nuclear engineers must take a more conservative approach in the design of their respective technological systems by assuming that some type of defect is present at the outset. In fact, the United States Air Force requires that the aircraft manufacturer guarantee a certain number of flying hours based on an initial assumption that a crack of some measur-

---

* The word "plastic" in the title of this treatise obviously refers to a class of materials. In the context of the footnoted sentence, we refer not to a "material" but rather to a state of deformation resulting from the application of a stress that exceeds the material's yield strength. By the term "plastic deformation," then we refer to deformation that is irrecoverable and, hence, accumulatable to the limiting capacity of a particular material. By contrast, "elastic deformation" refers to a reversible state of deformation wherein removal of all body forces returns the object to its original shape and size without memory of its prior load history.

able length exists at *every* rivet hole in the plane [1]. The position taken by a design engineer in any given situation will depend upon prior field experience, relative to the probability of encountering adventitious defects, and the potentially dangerous consequences of an unexpected early life failure.

For convenience, the authors have opted to treat separately the respective literatures dealing with unnotched (Chapter 2) and notched (Chapter 3) test bar experimental results. However, points of commonality and dichotomy shall be emphasized.

2.2   FATIGUE TEST VARIABLES

Assuming for the moment that one may wish to predict the total fatigue life of some component, certain laboratory test results are needed. This, in turn, requires that a specimen geometry be chosen, certain control variables (e.g., stress or strain) be established, and various test conditions chosen (e.g., test temperature, cyclic frequency, loading rate and environment) so as to simulate as well as possible actual component service conditions. Alternatively, one may wish to consider the conditions associated with a particular service failure that might involve laboratory tests conducted under simulated conditions. For example, consider the fatigue-induced fracture of the plastic tricycle shown in Fig. 2.1. Let us assess the variables cited that may have been associated with this service failure. One may inquire how large the children were who rode the tricycle. How often did each child (presumably of somewhat different weights) scoot about, and what was the duration of each ride? How did they seat themselves, by alighting gently or by flopping into the seat? Was the tricycle cared for, or thrown about, thereby subjecting its parts to potential damage? Did the children ride the tricycle more during the summer than in the cooler fall and winter months? In this regard, what was the character of the annual riding histogram? Did the children park the bike in a shelter or was it exposed to considerable ultraviolet irradiation as well as various forms of precipitation? The reader can surely add to the list of possible variables that may have contributed to the eventual fracture. While some of the above factors may have been of secondary importance, others will have had, in themselves, a major impact on the fracture process. Finally, some of these variables may have interacted synergistically to effect a major influence of the tricycle material's resistance to cyclic loading.

Though fracture of a plastic tricycle may not send shock waves through the commercial world of plastic component manufacturers, this failure may

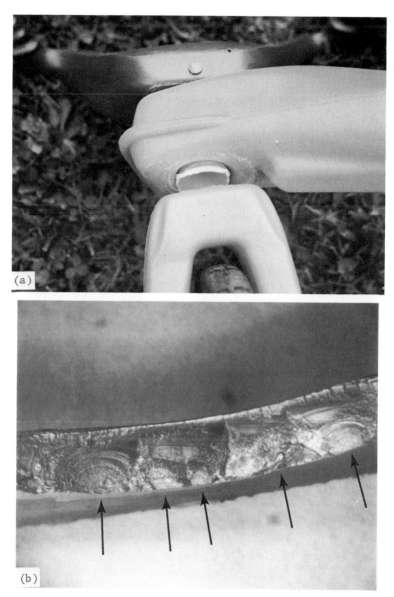

Fig. 2.1 Fatigue fracture of plastic tricycle. (a) Broken assembly; (b) arrows indicate several origins of fatigue cracks. Note "clam-shell" markings emanating from each origin which represent periods of crack growth during life of component. (Courtesy Jason and Michelle Hertzberg.)

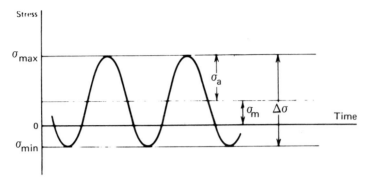

Fig. 2.2   Key stress variables associated with cyclic loading.

well be symptomatic of potential fracture of more significant engineering structures. Consequently, the relevant loading and environmental parameters responsible for one failure may, indeed, be representative of a general class of fractures. To identify these important fracture variables is then of primary interest to both manufacturer and consumer.

Key stress variables that contribute in different ways to the fatigue process under stress-controlled test conditions are shown in Fig. 2.2 and are defined by

$$\sigma_{max}, \sigma_{min} = \text{maximum and minimum stresses, respectively,}$$
$$\Delta\sigma = \text{stress range} = \sigma_{max} - \sigma_{min},$$
$$\sigma_a = \text{stress amplitude} = \tfrac{1}{2}(\sigma_{max} - \sigma_{min}), \qquad (2.1)$$
$$\sigma_m = \text{mean stress} = \tfrac{1}{2}(\sigma_{max} + \sigma_{min}),$$
$$R = \text{stress ratio} = \sigma_{min}/\sigma_{max}.$$

In related fashion, one may choose to define other quantities based on the variation of strain with time for the case of strain-controlled testing. For convenience, discussion of stress- and strain-controlled fatigue testing will be considered separately in this chapter.

2.3   STRESS-CONTROLLED TESTING

Researchers have found that the magnitude of the alternating stress or stress range has a strong bearing on the fatigue life of a component or test specimen. For this reason, most fatigue data in the plastics and metals industries compare cyclic life as a function of stress amplitude. Figure 2.3 provides two examples of the usual manner by which fatigue data are presented [1a]. These stress–log cyclic life plots (often referred to as S–N dia-

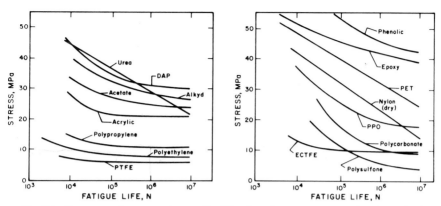

Fig. 2.3 Representative stress–log cyclic life plots for numerous engineering plastics [Riddell (1a)].

grams) reveal several important facts. First, the cyclic life of any material increases with decreasing stress amplitude. Furthermore, below a certain stress level, some materials seem to possess "infinite" life to the extent that failure does not occur in less than 10^7 cycles.

2.3.1 Specimen Geometry and Loading Method

To generate these data in the laboratory, a number of different loading methods and specimen configurations have been developed. Loads can be introduced either by rotational bending, reciprocal bending, reciprocal torsion, or by pulsating axial loads. In rotating bending, one can effect either a uniform moment across the gauge section (Fig. 2.4a) or a bending moment that increases with increasing distance from the applied load point (Fig. 2.4b). The latter condition forces the fatigue fracture to initiate at the base of the fillet at the end of the gauge section; as a result, fatigue data from this specimen are found to depend strongly on the fillet geometry over and above material considerations. The use of a triangular specimen form associated with a reciprocating bending moment avoids this problem and provides for a uniform flexural stress across the entire gauge section. Indeed, this specimen shape and method of loading has been adopted by ASTM to serve as the recommended standard (ASTM Standard D671-71) by which the fatigue response of engineering plastics may be evaluated [2]. Two specimen shapes have been approved by this organization (Fig. 2.5), both providing for a constant bending stress across the gauge section.

One must be mindful of certain qualifications associated with this standard test method. Specifically, ". . . The results are suitable for direct application in design only when all design factors including magnitude and mode

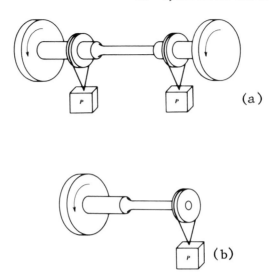

Fig. 2.4 Rotating bending fatigue tests. (a) Uniform bending moment; (b) nonuniform bending moment.

of stress, size and shape of part, ambient and part temperature, heat transfer conditions, cyclic frequency, and environmental conditions are comparable to the test conditions" [2]. The essential point to be recognized here is that specimen "failure" may take place by either of two modes. In one instance, a specimen may fail by the nucleation and growth of a crack across the gauge section. The fatigue life of such a sample would then be defined simply as the number of loading cycles endured prior to final specimen separation. By sharp contrast, some materials may become overheated as a result of the accumulation of hysteretic energy generated during each loading cycle. Since this energy is dissipated largely in the form of heat, an associated temperature rise will occur for every loading cycle when isothermal conditions are not met. In concert with such a specimen temperature rise, the elastic modulus is found to decrease. As a result, larger specimen deflections are required to allow for continuation of the test under constant stress conditions. In turn, these larger deflections will contribute to even greater hysteretic energy losses with further cycling, thereby producing an auto-accelerating tendency for rapid specimen heating. A point is reached whereby the specimen is no longer capable of supporting the loads introduced by the test machine within the deflection limits of the test apparatus. In such situations, ASTM Standard D671-71 defines fatigue failure life (thermal fatigue, in this instance) as the number of loading cycles at a given applied stress range that brings about an "... apparent modulus decay to 70% of the

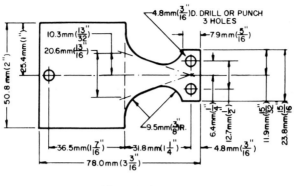

TYPE A

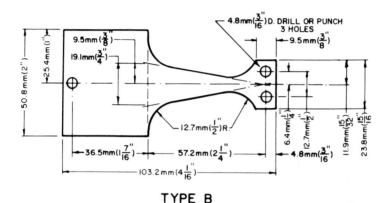

TYPE B

Fig. 2.5 ASTM Specification D-671 Constant Force Fatigue Specimens. [Reprinted with permission from the American Society for Testing and Materials, 1916 Race St., Philadelphia, PA 19103 © (2).]

original modulus of the specimen determined at the start of the test" [2]. In related fashion, the ASTM Standard calls for the investigator to "...measure the temperature at failure unless it can be shown that the heat rise is insignificant for the specific material and test condition" [2].

Since the tendency toward thermal fatigue failure depends strongly on test frequency (see Section 2.3.2), one should choose to conduct laboratory tests ideally at frequencies comparable to those anticipated for a component in service. As a rule, this is not done because the test apparatus described in Standard D671-71 is designed to operate at 30 Hz only and would require changes in the drive unit and spring-mass system to achieve other frequencies.

Other test procedures have been proposed, based on constant specimen deflection, which avoid the problem of specimen heating and will be discussed in Section 2.4.

2.3.2   Fatigue Failure Resulting from Hysteretic Heating

During the past four decades, considerable attention has been given to the nature of hysteretic-heating-induced thermal fatigue failures in rubbers and engineering plastics. Earlier studies pointed to the important role of applied stress, test frequency, specimen dimensions, heat transfer characteristics of the specimen and material, ambient temperature, and material properties in establishing test conditions for thermal failure [3–15]. The major cause of thermal failure is generally believed to involve the accumulation of hysteretic energy generated during each loading cycle. Since this energy is largely dissipated in the form of heat, the test bar will experience an associated temperature rise for every loading cycle when isothermal conditions are not met. The temperature rise can be so great as to cause the sample to melt, thereby preventing it from carrying any load. Note that such "failures" do not require that the specimen break into two pieces. Failure is thus presumed to occur by viscous flow though perhaps the occurrence of some bond breakage cannot be excluded.

Attempts have been made to quantify the rate of heat generation and temperature increase associated with each loading cycle [7, 9, 12, 16]. For the case of stress-controlled testing, Schmidt and Marlies found the rate of heat generation H_F to be given by [16]

$$H_F \propto DF^2/E_d,   \tag{2.2}$$

where H_F is the heat generation rate under force control, D the damping capacity of the material, F the force, and E_d the dynamic modulus. When heating takes place during the test, the dynamic modulus is seen to decrease. This, in turn, results in a greater degree of heat generation under constant loading conditions. Also, the degree of damping increases, which leads to further heat generation. Ratner and Korobov [7] showed the specimen temperature increase ΔT at equilibrium to be

$$\Delta T \propto \sigma^2 f d^2 D/E_d A,   \tag{2.3}$$

where σ is the stress amplitude, f the test frequency, d the specimen diameter, D the damping capacity of material, E_d the dynamic modulus, and A the heat transfer parameter. Here, again, the autoacceleration of specimen temperature rise with repeated load cycling is attributed to increased damping D and a decrease in dynamic modulus E_d. According to Ferry [17], the energy dissipation rate also may be described by

$$\dot{E} = \pi f J''(f, T)\sigma^2, \tag{2.4}$$

where $\dot{E}$ is the energy dissipation rate per unit time, f the frequency, J'' the loss compliance, and σ the peak stress.

Equation (2.4) may be reduced to show the temperature rise per unit time as noted in Eq. (2.5).

$$\Delta\dot{T} = \pi f J''(f, T)\sigma^2/\rho c_p, \tag{2.5}$$

where $\Delta\dot{T}$ is the temperature change/unit time, ρ the density, and c_p the specific heat. Note the strong similarity between Eqs. (2.3) and (2.5). The temperature rise for a given load cycle, based on these relationships, is not absolutely correct since the simplified relationships do not account for heat losses to the surrounding environment [13]. However, they are useful in that they identify the major variables associated with hysteretic heating. For example, constant force cantilever-type smooth specimens cycled at 30 Hz were used to evaluate the fatigue response of polytetrafluoroethylene (PTFE), a highly crystalline polymer, and resulted in numerous thermal failures [13, 14] (Fig. 2.6). Note the superposition of temperature rise curves corresponding to the various stress levels. It is seen that for all stress levels above the endurance limit (~ 6 MPa) (the endurance limit is considered to be the stress level below which fatigue failure is not observed), the polymer heated up to the point of melting as shown in the temperature rise curves A, B, C, D, and E. Evidently, heat generation was greater than heat dissipation to the surrounding. When a stress level less than the endurance limit was applied, specimen temperature became stabilized at an intermediate level below the point where thermal failure was observed. These specimens did not fail after 10^7 cycles. Furthermore, the rate of temperature rise was found to be closely related to the change in loss compliance J'', which is itself a function of temperature (Fig. 2.7). These data show that the loss compliance rises rapidly in the vicinity of the glass transition temperature after a relatively small change at lower temperatures. Consequently, it would be expected that the temperature rise in the sample would be moderate during the early stages of fatigue cycling but markedly greater near the final failure time. One may conclude, therefore, that thermal failure describes an event primarily related to the lattermost stages of cyclic life. Also shown in Fig. 2.7 is a representative curve indicating the temperature rise for a specimen tested at a stress level below the endurance limit. It is seen again, as compared to Fig. 2.6, that a limiting temperature for the specimen is reached. Note that this maximum temperature is associated with a relatively

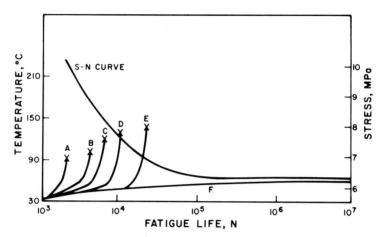

Fig. 2.6 Stress–log cyclic life data for polytetrafluorethylene. Temperature rise during fatigue tests also shown. × denotes fatigue failure, test frequency 30 Hz [Riddell, Koo, and O'Toole (13), with permission from Engineering Properties Laboratory, Plastics Division, Allied Chemical Corp.].

	Stress (MPa)	N (Cycles)	T (°C)
A	10.3	2×10^3	100
B	9.0	4×10^3	115
C	8.3	6.1×10^3	125
D	7.6	9.5×10^3	130
E	6.9	1.9×10^4	141
F	6.3	10^7	60

low value of J'' as compared to the value associated with the temperature where thermal failure occurs.

The preceding discussion was simplified by the assumption of a monotonically increasing level of J'' with increasing temperature. It is conceivable and even likely, however, that an intermediate temperature range would exist wherein J'' and the corresponding rate of hysteretic energy dissipation would decrease, thereby retarding the rate of specimen temperature elevation; such would be the case between the principal damping peaks as shown in Fig. 2.7. Obviously, the depth and position of the trough between the peaks will vary with polymer chemistry and architecture so that the region of attenuated temperature rise would be unique for each material.

Equations (2.3) and (2.5) also show that the fatigue life of a sample should decrease with increasing frequency since the temperature rise per loading cycle is seen to be proportional to the frequency (ignoring any interactive

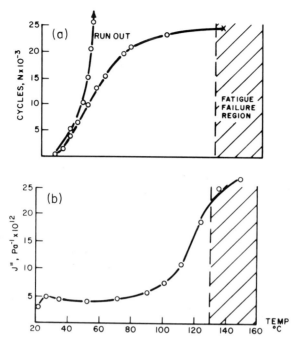

Fig. 2.7 Relationship between cyclic-induced specimen (a) temperature rise and (b) temperature dependence of loss compliance of polytetrafluorethylene [Koo, Riddell, and O'Toole (14), with permission from Engineering Properties Laboratory, Plastics Division, Allied Chemical Corp.].

effects of frequency [17] on J'', for example). The data shown in Fig. 2.8 clearly support the anticipated effect of this variable. Specimen configuration represents another variable that could affect polymer fatigue results. As mentioned above, the temperature rise for each loading cycle is dependent upon the amount of heat that is dissipated to the surroundings. Consequently, the fatigue life of a given sample should depend on the heat transfer characteristics of the sample and be enhanced by an increase in specimen surface area-to-volume ratio. In confirmation, it has been shown that the endurance limit increases with decreasing specimen thickness in PTFE, polypropylene (PP), and poly(methyl methacrylate) (PMMA) [7, 13]. In related fashion, the fatigue life of a specimen subjected to reversed bending should be greater than that associated with a sample uniaxially loaded [4, 18]. In the latter case, the entire specimen cross section experiences the maximum stress, whereas only the outermost fibers of the bend bar receive maximum loading. Hence, the bend sample gets hot at the surface only where heat dissipation can be achieved readily; the axial sample gets hot throughout the gauge area and becomes subject to thermal failure at a lower stress level.

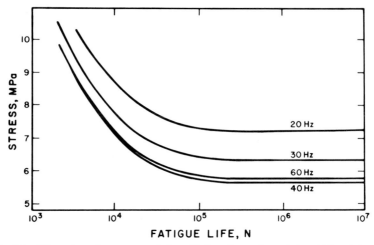

Fig. 2.8 Effect of test frequency on fatigue life in polytetrafluorethylene [Riddell, Koo, and O'Toole (13), with permission from Engineering Properties Laboratory, Plastics Division, Allied Chemical Corp.].

(A similar influence of stress gradient is found for test conditions leading to failure by fatigue crack propagation processes.) Thus, it is apparent that specimen shape and size and stress gradient will affect test results. This constitutes a major drawback to unnotched specimen tests subject to thermal failure since the test results will be a function of specimen geometry and, therefore, not reflect the intrinsic response of the material being evaluated.

Returning to the observed fact that most of the temperature rise occurs during the latter stages of the fatigue life, other tests have been reported wherein intermittent rest periods were interjected into the cyclic history of the sample. In this way any temperature rise in progress could be dissipated during the rest periods. As such, one would expect the fatigue life of speci-mens allowed intermittent rest periods to be substantially greater than un-interrupted test samples. Indeed, significant improvements in fatigue life have been noted when intermittent rest periods were introduced during the test [13, 19–21]. On the basis of these results, Koo *et al.* [14] concluded that the concept of cumulative damage is inoperative for conditions involving thermal failures. In relative fashion, Prevorsek *et al.* [22] restricted their linear rule of cumulative damage to situations where ". . . the temperature of the specimen and of the zone surrounding the propagating crack front do not vary with experimental conditions."

With a carefully chosen program of rest periods, an infinite fatigue life would be expected so long as the sample temperature rise is never allowed to exceed the limiting steady-state range associated with the endurance limit and a crack does not nucleate. While this reinforces the contention that thermal failure is primarily associated with a rapid temperature rise occurring during the latter stages of fatigue life, the use of fatigue data based on periodic rest period testing are certainly not appropriate for the case of polymeric structures subjected to continuous loading cycles.

Proceeding further, Cessna *et al.* [15] conducted isothermal tests of unnotched samples by introducing a temperature measuring and controlling closed loop device into the fatigue apparatus. In a comparative study of glass-reinforced polypropylene, these authors showed a 20–100-fold increase in fatigue life of this material under isothermal testing conditions, as compared to the nonisothermal condition where the specimen temperature was allowed to rise freely. A marked retention of original specimen stiffness for isothermal testing as compared to nonisothermal testing was noted. Cessna *et al.* [15] concluded that mechanically induced fatigue failures in the materials examined (occurring at very low-frequency fatigue testing condition) could be simulated by *isothermal* high-frequency fatigue tests so as to avoid the thermal failure mechanism. Broutman and Gaggar [19] reported similar effects in their fatigue study of epoxy and polyester resins. An increase in fatigue resistance was noted in specimens air cooled during testing and in specimens immersed in circulating water. In the latter instance, the water served to draw the heat away from the cyclically loaded sample. In this connection, Ratner and Barash [6] concluded that the beneficial effect of petroleum ether, water, and alcohol atmospheres present during the fatigue testing of an unswollen polyamide was due to cooling of the test bar resulting from evaporation of these volatile species at its surface. By comparison, fatigue life was not enhanced when tests were conducted in nonvolatile liquid mediums such as glycerin and oil.

It is appropriate at this point to consider the consequence of cyclically loading a polymeric solid at a stress level below that necessary to cause thermal runaway, i.e., at a stress level where the specimen temperature reaches a steady-state value below the glass transition temperature T_g or the crystalline melting temperature T_m. When such a test is conducted, it is possible for the sample to fail anyway, this time by a mechanical process involving fatigue crack initiation and propagation. As such, designing a component so that it is not subject to thermal failure does not also preclude the possibility for a mechanical-type fatigue fracture. Herein lies another major shortcoming of the thermal failure theory for fatigue of polymers. The results of Crawford and Benham [18] are especially illuminating in this regard. Figure 2.9 reveals the temperature rise during axial cycling in polyacetal as a function of stress level. Compare the tendency toward increased specimen temperature at increasing stress level with the results shown in

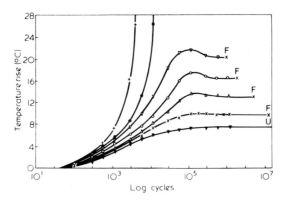

Fig. 2.9   Temperature increase resulting from uniaxial cycling at 5 Hz with load control, sine wave and zero mean stress in polyacetal. ▼, 15MPa; ×, 16 MPa; △, 17.4 MPa; ○, 19.7 MPa; ▽, 21.6 MPa; ●, 22.4 MPa; +, 27.8 MPa; F = fracture; U = unbroken. [Reproduced from R. J. Crawford and P. P. Benham, *Polymer*, 1975, **16**, 908, by permission of the publishers, IPC Business Press Ltd. ⓒ.]

Fig. 2.6. Also note that some samples failed by mechanical processes and others by melting. This is clearly evident from the results shown in Fig. 2.10. The curve at the right corresponds to mechanical failures in numerous polyacetal samples cycled in reversed loading at frequencies from 0.167 to 10 Hz; these failures are found not to be frequency sensitive. In marked contrast, the thermal failures depended strongly on test frequency. The

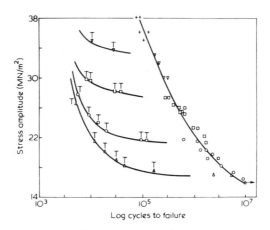

Fig. 2.10   Stress–log cyclic life in polyacetal. Note distinction between mechanical and thermal failures *T*. +, 0.167 Hz; ▽, 0.5 Hz; □, 1.67; ○, 5 Hz; △, 10 Hz. [Reproduced from R. J. Crawford and P. P. Benham, *Polymer*, 1975, **16**, 908, by permission of the publishers, IPC Business Press Ltd. ⓒ.]

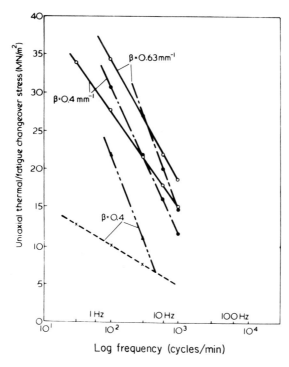

Fig. 2.11   Change-over stress from thermal to mechanical failure as a function of test frequency and specimen surface/volume ratio, β (mm^{-1}). Load cycle control, sine wave, zero mean stress. $\bigcirc$, Acetal; $\bullet$, PMMA; $\times$, PP; $\blacktriangle$, unfilled PTMT. [Reproduced from R. J. Crawford and P. P. Benham, *Polymer*, 1975, **16**, 908, by permission of the publishers, IPC Business Press Ltd. ©.]

transition from thermal to mechanical failure was described by Crawford and Benham in terms of a "change-over stress level," which was found to depend on test frequency, mean stress, cyclic waveform, and specimen surface area-to-volume ratio (see also Section 5.1.5). The relationship between the uniaxial change-over stress and test frequency is seen in Fig. 2.11 and described empirically by

$$\sigma_{\mathrm{c}} = [A/V]^{1/2}\{C_1 - C_2 \log_{10} f\}, \qquad (2.6)$$

where σ_{c} is the change-over stress, A the specimen surface area, V the specimen volume, and f the cyclic frequency.

A further treatment of polymer fatigue was provided by Opp, Skinner, and Wiktorek [23]. Based upon a hysteresis energy model [24], Opp *et al.* developed a relationship that showed total hysteresis energy per cycle to be the sum of both mechanical energy and thermal energy. For a given set of testing conditions, either a thermal failure or mechanical failure could be predicted from the model. As such, their model represents a bridge between thermal- and mechanical-type failures by predicting thermal failures under high-cyclic stress conditions and mechanical failures at low-cyclic stress levels. In addition, their model predicts that fatigue life should increase with decreasing frequency, increasing surface area-to-volume ratio, and with the imposition of periodic rest periods so as to curtail temperature rise; compare this with results discussed above [e.g., Eqs. (2.3), (2.5), and (2.6) and Figs. 2.8–2.11]. Their solution also suggests that fatigue life should be a function of the waveform of the cyclic loading. Indeed, waveform effects on fatigue crack propagation in notched specimens of homogeneous polymers have been observed and are discussed in Section 3.4. Unnotched composites are discussed in Section 5.4.

Using a heat dissipation argument, Opp *et al.* [23] concluded that the presence of a stress concentration may *increase* fatigue life. Their position was based upon the notion that since stress concentrations are usually found on the surface, hysteretic heat would be dissipated more readily, thereby restricting a temperature rise. While such a postulate appears plausible within the limited confines of a thermal heat transfer argument, it is certainly suspect in a more general sense. For example, Gotham [25] reported decreasing fatigue strength with decreasing radius of the notch root (Fig. 2.12).

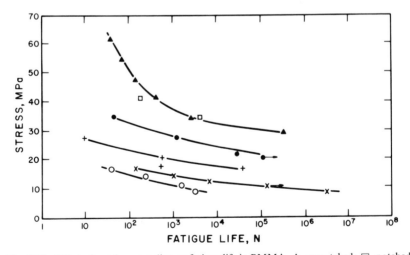

Fig. 2.12 Effect of notch root radius on fatigue life in PMMA. ▲, unnotched; □, notched, $r = 2$ mm; ● notched, $r = 1$ mm; +, notched, $r = 500$ μm; ×, notched, $r = 250$ μm; ○, notched, $r \approx 10$ μm; uniaxial tension/compression, 0.5 Hz, 20°C, 65% relative humidity, $\sigma_{mean} = 0$ [Gotham (25)].

Crawford and Benham [18] also found the fatigue strength of notched polyacetal samples to be lower than smooth ones, though the difference was not as great as that expected. They argued that the localized volume of highest stress at the notch root would generate less heat so that ". . . higher frequencies can be operated [in notched samples] for the same peak stress compared with the unnotched specimens [18]. To this extent, these results recognize thermal effects, but the overwhelming effect of the notch on fatigue life proved to be negative. As Andrews [26] stated in his review of fatigue in polymers, "It is obvious that a specimen containing stress concentrations (e.g., at notches or holes) will fail more readily than one which is uniformly stressed." The danger in application of thermal arguments alone in the explanation of fatigue failure in polymers is, thus, clearly evident.

A paper by Constable, Williams, and Burns [27] serves as the junction point from which mechanical aspects of fatigue have received considerable attention. Fatigue failure in PMMA and PVC was considered to occur by crack propagation from existing defects in the material so long as this event was not preceded by thermal failure. Though the bimodal failure mechanism concept is similar to that proposed by Opp *et al.* [23], it does differ in that the mechanical-type failure is specifically identified with fatigue crack propagation as opposed to the accumulation of a certain level of hysteresis energy as proposed by Opp *et al.* It is important to note at this point that to these authors' knowledge, no thermal failures have ever been observed in fatigue crack propagation (FCP) tests of precracked samples. In view of the fact that the applied cross-section stress (and strain) range is below the unnotched specimen endurance limit, the absence of heat buildup and consequent thermal failure is not surprising (e.g., see Figs. 2.6 and 2.9). However, there still does exist the possibility that stable crack growth across the sample may consist of small-scale (the order of the crack tip plastic zone size) local melting that propagates across the sample. Since "thermal" failures never really produce fracture surfaces, it is not possible to compare fracture morphological features in the two cases. However, a comparison of fatigue markings produced by stable crack extension at different frequencies (see Chapter 4) discredits the concept of stable crack advance by progressive crack tip melting. While hysteretic heat dissipation per cycle is undoubtedly very large within the crack tip plastic zone due to stress and strain amplification, it is believed that heat dissipation usually takes place from this small concentrated heat sink to the much cooler surrounding material and serves to preclude an uncontrolled temperature rise. To be sure, crack tip temperature rises have been recorded in PVC, PMMA, and PC by Attermo and Östberg [28], which are in agreement with predictions by Barenblatt *et al.* [29], but which do not cause local melting. More recent studies of certain impact-modified nylon 66 blends have shown considerable crack tip temperature elevations that did not cause thermal melting but did have an adverse effect on the material's resistance to fatigue crack growth (see Section 3.4). One key to the thermal fatigue mechanism clearly relates to size effects [13, 14, 18, 25] with thermal failures being suppressed by an increase in the surface area/volume ratio of the sample. In addition, the likelihood of thermal failure should depend on the ratio of material volume experiencing maximum hysteretic heating as compared to the total volume of the test sample. When the relative volume of heated material is high, as in the case of an unnotched sample experiencing large stress (or strain) excursions across the entire gauge section, thermal failure will occur. Conversely, hysteretic heating restricted to a crack tip should not produce a thermal failure.

2.4 DEFLECTION-CONTROLLED TESTING

Constant force or stress amplitude fatigue testing of unnotched and notched specimens is appropriate for the evaluation of materials chosen for components that will be subjected to load-controlled conditions. Examples of such components include rotating shafts, pressure vessels that experience cyclic internal pressures, and rods or ropes that must withstand fluctuating axial loads. And yet, other situations exist wherein some components become loaded within certain limits of deflections or extensions. Various coil and leaf springs, shoe soles, and components subjected to thermal cycling represents examples of components loaded in this manner. Furthermore, it is possible to identify components that experience both load and displacement control. For example, the side walls of a pneumatic tire during road use experience essentially constant amplitude of deflection while the tire tread portion in contact with the road surface experiences a constant force compression cycle. It is appropriate, therefore, to examine the major aspects of constant deflection or strain amplitude testing and to determine whether the relative ranking of material fatigue performance changes when the principal test variable is altered from load to deflection control.

A proposed method of test involving constant amplitude of deflection has been suggested by ASTM [30]. The test machine is of the fixed-cantilever, repeated constant-deflection type. It is found that the initial stress range induced by the applied deflection limits decays with repeated cycling as shown in Fig. 2.13. This attenuation in specimen stress range in response to the fixed displacements is caused by plastic deformation and hysteretic-heat-

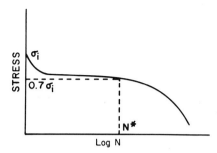

Fig. 2.13 Stress decay curve. Fatigue life N^* corresponds to cyclic stress level equal to 70% of initial value σ_i.

induced softening of the material. Cyclic-induced material property changes may also occur and are discussed more fully in Section 2.5. When the resulting stress range decreases, it has been suggested that the fatigue life be defined when the instantaneous stress level decreases to 70% of its initial value. Note the similarity between this arbitrary failure criterion and the 70% elastic modulus value used in constant load amplitude testing (see Section 2.3.1). When the change in stress during the test is not monitored, fatigue data portrayed in the form of a S–N diagram should be clearly labeled to show the relationship between fatigue life and the *initial* stress range at the beginning of the test. In fact, it has been argued that stresses should not be calculated from deflection-controlled test conditions since the dynamic modulus changes during the initial portion of the test, thereby altering the strength-of-materials relationship between specimen deflection and induced stress level.

Since the applied stress decreases with cycling under conditions of constant specimen deflection, one finds a decreased tendency toward hysteretic heating. As a result, constant-deflection testing does not lead to thermal failures. Schmidt and Marlies [16] showed that the heat generation rate H_x under fixed displacement conditions can be described by

$$H_x \propto DE_d x^2, \tag{2.7}$$

where D is the damping capacity, E_d the dynamic modulus, and x the displacement amplitude. During initial cycling, hysteretic heating leads to a decrease in E_d which, in turn, acts to reduce the heat generation rate. Therefore, subsequent specimen heating will be lower and a steady-state temperature quickly reached. An improved constant deflection test method has been suggested by Gauvin and Trotignon, which not only avoids thermal heating failures but allows for simultaneous testing of up to 16 specimens [31].

The principal point to recognize here is that the relative ranking of fatigue behavior among various polymers may well depend upon the nature of the fatigue test itself. For example, a polymer that exhibits considerable damping may possess a low fatigue limit under constant stress amplitude testing conditions due to premature thermal melting; without such uncontrolled heating as would exist in constant-deflection amplitude testing, this material might well exhibit a higher ranking relative to the other materials. Furthermore, when constant stress tests are conducted, the relative fatigue ranking of different materials will depend upon the extent to which the test is conducted under isothermal or adiabatic conditions. Therefore, it is extremely important for the design engineer to identify the service conditions of the component under study so that tests could be designed under the most relevant conditions.

2.5 STRAIN-CONTROLLED TESTING

A close connection exists between constant-deflection and constant-strain testing. Of particular importance, neither test condition leads to runaway creep of the sample or thermal failure. Strain-controlled fatigue tests are thought to be useful in evaluating the cyclic life of components that contain blunt notches and other moderate stress concentrations. This is due to the fact that the notch root region experiences a strain-controlled

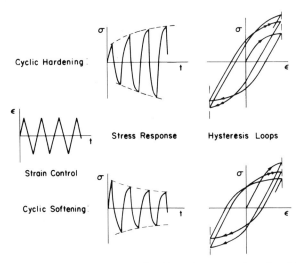

Fig. 2.14 Cyclic hardening and cyclic softening under reversed strains. [Reprinted from P. Beardmore and S. Rabinowitz, "Treatise on Materials Science and Technology, Vol. 6," R. J. Arsenault, ed., Academic Press, New York, 1975, p. 267, with permission.]

condition dictated by the much larger surrounding mass of essentially elastic material.

Most constant strain amplitude tests are conducted under fully reversed tension–compression loading conditions (i.e., push–pull mode). During the course of such testing, fundamental changes arise with regard to material response. From Fig. 2.14 we see that the height of the hysteresis loops may either increase or decrease, which reflects cyclic strain hardening and softening, respectively. For metals, as a general rule, initially hard materials will cyclically soften while soft materials will cyclically harden [32].

Though the extant literature regarding cyclic strain fatigue of metal alloys is extensive, relatively few studies, notably those by Beardmore and Rabinowitz [33–36] and Tomkins and Biggs [37–39], have been performed for polymers. In a recent review of the polymer cyclic strain literature, Beardmore and Rabinowitz reported that only cyclic strain softening occurs in crystalline, amorphous, and composite polymers [36]. The change in cyclic stress amplitude associated with constant strain amplitude test conditions is shown in Fig. 2.15 for several materials. Though differences exist with respect to the specific shape of these data plots, the overall trend of cyclic softening is clearly evident. Four distinct regions are identified for the case of PC (Fig. 2.15a). The incubation stage is believed to be associated with an initial material state involving few mobile defects and a state wherein such defects are generated at a slow rate. (The movement and multiplication of these defects is related closely to the cyclic softening process.)

Tests have shown that the duration of this first stage depends upon both the initial state of the material and the imposed strain amplitude. In materials that are structurally inhomogeneous, such as in semicrystalline and two-phase polymers and composites, more initial defects are likely to exist, which accounts for the absence of the initiation stage in nylon and ABS (Figs. 2.15b and 2.15c). Also, when the imposed strain amplitude is increased, the extent of the incubation stage decreases. After a transition region, wherein the material softens appreciably over a short period of cycling, the polymer then enters a third stage associated with a relatively steady-state condition between applied strain and induced stress ranges. This behavior has been observed in PC (Fig. 2.15a), nylon (Fig. 2.15b), polystyrene, poly(methyl methacrylate), polypropylene, and glass-reinforced nylon [33–36] (see also Section 5.3.1). ABS polymer differs, however, in that a stable steady-state region is never achieved (Fig. 2.15c); instead, the hysteresis loop heights continually decrease with cycling (see also Section 5.1.5). A steady-state condition implies a dynamic balance between structural hardening and softening of the material. In addition, polymeric solids also undergo molecular orientation hardening, involving realignment of individual molecules and their entanglements. In this regard, it has been found that PC ex-

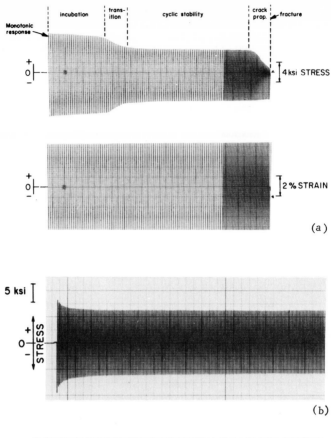

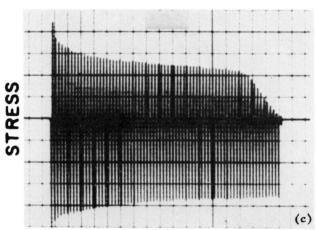

Fig. 2.15 Examples of cyclic strain softening at 298°K in (a) polycarbonate, (b) nylon 66, (c) ABS. [Reprinted from P. Beardmore and S. Rabinowitz," Treatise on Materials Science and Technology, Vol. 6," R. J. Arsenault, ed., Academic Press, New York, 1975, p. 267, with permission.]

periences a 1% increase in density in the cyclic stable state, and nylon exhibits similar behavior [36]. Surely some structural rearrangement in these two polymers must have occurred to account for the observed decrease in internal free volume.

It is important to determine the cyclically stabilized stress–strain response of a material since it may be quite different from the initial monotonic response. For the case of cyclic strain, the fatigue and overall mechanical response of a given engineering component will be more accurately predicted from a cyclically stabilized stress–strain curve than from its monotonic counterpart. Cyclically stabilized stress–strain curves may be obtained in several ways. In one approach, a series of companion samples are cycled within various strain limits until the respective hysteresis loops become stabilized. The cyclic stress–strain curve is then determined by fitting a curve through the tips of the various superimposed hysteresis loops as shown in Fig. 2.16 [40]. An even quicker technique involving only one sample has been found to provide excellent results and is used extensively in current cyclic strain testing experiments. As seen in Fig. 2.17, the specimen is sub-

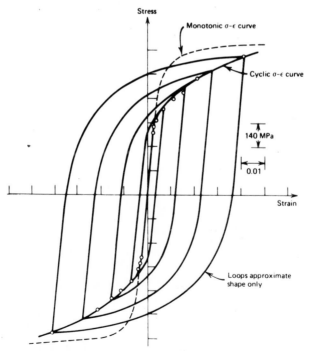

Fig. 2.16 Determination of cyclic stress–strain curve from consecutive hysteresis loops. Monotonic stress–strain curve included for comparison. [Reprinted with permission from the American Society for Testing and Materials, 1916 Race St., Philadelphia, PA 19103 © (40).]

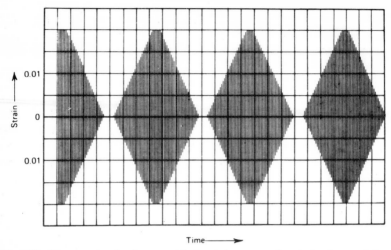

Fig. 2.17 Diagram showing incremental step test program for strain–time variations to achieve cyclic stress–strain curve.

jected to a series of blocks of gradually increasing and then decreasing strain excursions [41]. After a relatively few such blocks, the material reaches a stabilized condition. The investigator then simply draws a line through the tips of each hysteresis loop, from the smallest strain range to the largest. The resulting stress–strain curve then represents an envelope of stabilized stress range values associated with the various applied strain ranges. By initiating the test with the maximum strain amplitude in the block, the monotonic stress–strain curve is automatically determined for subsequent comparison with the cyclically stabilized curve. Consequently, both the monotonic and cyclic stress–strain curves can be determined from the same specimen.

A comparison of cyclic and monotonic stress–strain curves for several polymers is shown in Fig. 2.18. Two important features are to be noted. First, all the materials, semicrystalline, amorphous, and two phase, exhibit pronounced cyclic strain softening. This is to be expected, based on the previous discussions pertaining to the hysteresis loop heights (see Fig. 2.15). Second, the yield point exhibited by nylon, polycarbonate, and ABS under monotonic loading is eliminated in the cyclically stabilized state. Design engineers should be aware of these cyclic strain-induced material changes since ignorance of this information would lead to erroneous assumptions of *effective* material properties.

Cyclic strain–fatigue life data can be portrayed in yet another manner. Coffin [42] and Manson [43] have demonstrated in metals that the low cycle

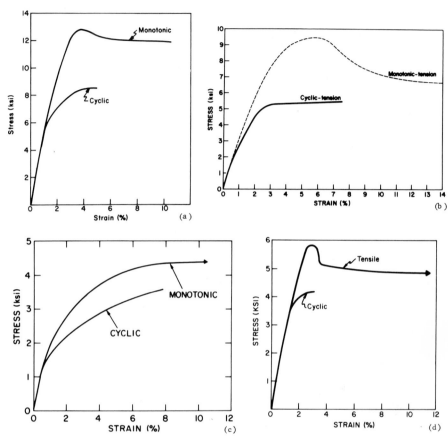

Fig. 2.18 Tensile cyclic and monotonic stress–strain curves at 298°K for (a) nylon 66, (b) polycarbonate, (c) polypropylene, (d) ABS. [Reprinted from P. Beardmore and S. Rabinowitz, "Treatise on Materials Science and Technology, Vol. 6," R. J. Arsenault, ed., Academic Press, New York, 1975, p. 267, with permission.]

fatigue life of smooth samples is related to the range of applied plastic strain with a relation of the form

$$\Delta \varepsilon_{p} \, N^{\alpha} = C, \qquad (2.8)$$

where $\Delta \varepsilon_{p}$ is the plastic strain range, N the cyclic life, α equals 0.5–0.7 for metals, and C is the constant, related to the true fracture strain in tension. Tomkins and Biggs [37] verified this relationship in nylon and observed $\alpha = 0.23$, which reflects a stronger dependence of strain range on cyclic life than in metals. Alternatively, one may show fatigue life data of a polymer as a function of the total strain amplitude, $\Delta \varepsilon_{T}$, which combines both elastic

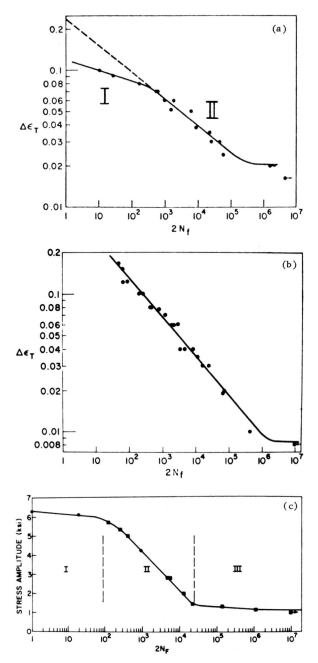

Fig. 2.19 Total strain amplitude versus fatigue life at 298°K in (a) PMMA, (b) polycarbonate, (c) cyclic stress–log fatigue life in polystyrene: ●, strain control; ■, stress control. [Reprinted from P. Beardmore and S. Rabinowitz, "Treatise on Materials Science and Technology, Vol. 6," R. J. Arsenault, ed., Academic Press, New York, 1975, p. 267, with permission.]

and inelastic strain components. This method of data presentation is analogous to the stress amplitude–fatigue life plots shown earlier (Fig. 2.3 and also shown in Fig. 2.19c). Typical plots, which reflect the two basic trends in polymeric response, are shown in Figs. 2.19a and 2.19b for PMMA and PC. The steep slope portion of these curves (stage II) conforms to an equation of the form given in Eq. (2.8) but modified by the term $\Delta\varepsilon_T$ in place of $\Delta\varepsilon_p$. Beardmore and Rabinowitz have found that the development of stage I depends primarily on the deformation behavior of the polymer [34–36]. When crazing occurs, a distinct change (lowering) occurs in the stage II slope. That is, the development of crazes during cyclic straining serves to lower the material fatigue life at high strain–short life levels. Similar behavior has been noted in polystyrene at room temperature, which is also known to craze under these conditions [33]. In fact, since the tendency to craze in PC is greatly enhanced at $78°K$, this material also exhibits stage I behavior under such low temperature test conditions [36].

2.6 CYCLIC LIFE DEPENDENCE ON MATERIAL, ENVIRONMENT, AND STRESS VARIABLES

The published literature regarding the effects of polymer chemistry and internal structure on the fatigue behavior of unnotched specimens is, indeed, limited. The most extensive studies have been reported in the Russian literature with studies by others, if conducted at all, apparently being confined to corporate proprietary literature. When examining the results of such investigations, it is of utmost importance to interpret the data trends in terms of the actual test procedures (i.e., deflection versus load amplitude control) employed to generate the data. For example, if tests were conducted under constant deflection conditions, then chemical and/or structural changes that lower polymer stiffness would be expected to lead to superior fatigue behavior. This comes as a result of a lower stress range being developed in the more compliant sample under constant deflection test conditions. Conversely, were the tests to be conducted within load amplitude limits, fatigue behavior would be expected to deteriorate since the more compliant samples would be strained more extensively for each loading cycle.

Bareishis and Stinskas [44] made a special effort to examine the effect of loading procedure in their study of the fatigue life sensitivity of polycaproamide to various stabilizers. These studies extended the work of Sogolova [45], Savkin *et al.* [46], and Kuchinskas and Machyulis [47] who found earlier that certain additives and antioxidants benefited the fatigue behavior of crystalline polymers. It was determined that such additives served to enhance spherulite nucleation while spherulite growth kinetics remained essentially unchanged; this, in turn, led to the development of a polymer internal structure composed of both finer and more uniform spherulites that exhibited superior fatigue properties. Bareishis and Stinskas found that the introduction of quinhydrone in specimen surface layers by diffusion stabilization contributed more toward improving fatigue properties than other stabilizers including molybdenum disulfide, iodine, and potassium iodide [44]. Again, this improvement was traced to a refinement in crystal structure. Since the latter contributed also to an increase in material stiffness, the relative ranking of these additives was examined by fatigue testing under both load- and strain-controlled conditions. The ranking of these additives was unchanged as a function of test procedure. It was noted, however, that a relatively larger beneficial effect of stabilizer addition on fatigue behavior was found in stress-controlled tests than in the strain-controlled ones. As discussed above, this was due to the fact that a smaller strain range was required in the stiffer material under load-controlled conditions, whereas, a larger stress range was necessary when the tests were conducted under cyclic strain conditions. Furthermore, Savkin *et al.* [46] noted that a finer and more uniform spherulite structure reduced the amount of hysteretic heating and the specimen temperature rise under constant stress amplitude test conditions.

Warty *et al.* [48] have demonstrated that the application of certain coatings can exert a large influence on the fatigue life of polystyrene test bars. They found that coating samples with a 600 MW polystyrene oligomer improved fatigue life at a given alternating stress by about an order of magnitude. This marked improvement was believed to be related to surface plasticization by this viscous liquid (T_g < ambient), which acted to blunt surface flaws, thereby reducing local stress concentrations. On the other hand, when polystyrene samples were coated with a material of 900 MW, the fatigue life decreased by 30%. In this instance, this coating, which was applied as a liquid at 50°C, cooled to form a brittle lacquer. Subsequent fatigue cycling then introduced cracks in this layer at an early stage so as to reduce the cyclic life of samples prepared in this manner.

The effect of modulus-induced changes on the fatigue behavior of polycarbonate and polyethylene has been studied more recently by Agamalyan [49]. By subjecting these two materials to gamma irradiation, he noted a distinct lowering of the strength of polycarbonate while the elastic modulus of polyethylene was seen to increase. Since polycarbonate tended to fail by mechanical processes rather than by thermal melting, irradiation exposure was seen to decrease the fatigue life of polycarbonate under constant stress amplitude test conditions. By contrast, polyethylene, which was reported to fail primarily by thermal melting, was found to exhibit a higher fatigue strength because of the irradiation-induced increase in elastic modulus [see Eqs. (2.2) and (2.3)].

The influence of test and heat treatment environment and polyamide moisture content has been studied to a limited extent. Stinskas *et al.* [50] reported that the beneficial effect of heating specimens in oil and/or water at 180 and 100°C was due to temperature-induced structural changes rather than material changes brought about by the surrounding environment. The most significant improvement in fatigue properties of polycaproamide (nylon 6) under stress-controlled test conditions was brought about by oil heating at 180°C, which produced a uniform spherulitic structure with monoclinic crystal packing. By contrast, the untreated material contained a less perfect spherulitic structure with a hexagonal crystal form and exhibited inferior fatigue properties (Fig. 2.20). The fatigue properties of polytetra-fluoroethylene were also found to depend upon heat treatment. Riddell *et al.* [13] found that the fatigue limit of this material decreased with increasing quench rate. Since the degree of crystallinity varied inversely with cooling rate, the inferior fatigue behavior of rapidly cooled specimens was related to a correspondingly higher amorphous fraction prone to enhanced hysteretic heating (Fig. 2.21). (For discussions of the effects of crystallinity and thermal history, see Sections 3.8.5 and 3.8.6, respectively.)

Strain-controlled testing has shown some interesting effects of plasticizer content and heat treatment on the fatigue behavior of several polymers. Van Gaut [51] reported that a greater plasticizer content increased the fatigue life of PVC. [Recall that under constant deflection testing, a lower modulus (associated with a greater plasticizer content) will involve a lower induced

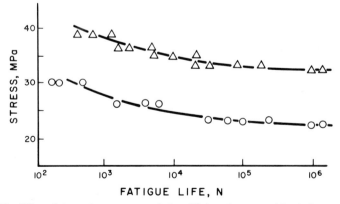

Fig. 2.20 Effect of thermal treatment on fatigue life in polycaproamide. △, heat treated for 1 hr at 180°C in oil; ○, as received. [Reprinted from Stinkas, Baushis, and Bareishis, with permission of Plenum from copyrighted material (50).]

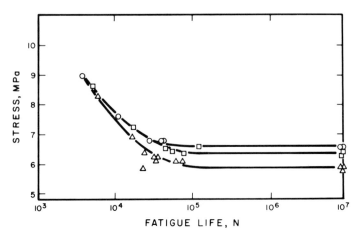

Fig. 2.21   Enhanced fatigue life (test frequency, 30 Hz) in polytetrafluorethylene with increasing crystallinity: △, low crystallinity, air quenched; □, medium crystallinity, 33.3°C/hr cooling; ○, high crystallinity, 5.6°C/hr cooling. [Riddell, Koo, and O'Toole (13), with permission from Engineering Properties Laboratory, Plastics Division, Allied Chemical Corp.].

stress range and hence lead to longer fatigue life.] On the other hand, heating PVC in an oven between 80 and 120°C for up to 50 hr brought about a sharp decrease in fatigue life as a result of an increase in the degree of cross-linking and associated stiffness.

In related fashion, Machyulis *et al.* found for the case of constant *deflection* testing that the fatigue properties of polyamides were improved with increasing moisture content [52]. However, when other tests were conducted under cyclic *load* conditions, the hysteretic heat buildup drove off moisture from the sample and tended to harden the material [53]. No hardening was observed with initially dry samples. Bareishis and Stinskas [53] also noticed that some of the hardening found in moisture-containing samples was related to cyclic stress-induced spherulite refinement in the material. Such evidence for structure hardening would appear to contradict the findings of Beardmore and Rabinowitz who concluded that only cyclic softening took place under cyclic strain testing [33–36]. Since the test procedures used by Bareishis and Stinskas were not clearly defined, these data must be evaluated with caution.

The effect of various organic liquid environments on the fatigue response in polystyrene was examined by Warty *et al.* [48]. Several alcohols, glycerol, and water were examined and the results shown in Fig. 2.22; *n*-butanol was found to have the most adverse effect of polystyrene fatigue life while water actually improved material response. Noting that the solubility parameter for polystyrene is 9.1 $(cal/cm^3)^{1/2}$ [54], these authors concluded that fatigue life decreased as the difference between the polystyrene and liquid environment solubility parameters approached zero. Under such conditions, the

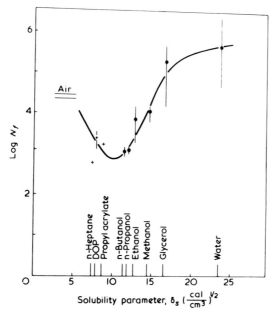

Fig. 2.22   Effect of solubility parameter on fatigue life in polystyrene. [Reproduced from S. Warty, J. A. Sauer and D. R. Morrow, *Polymer*, 1978, **19**(2), 1465, by permission of the publishers, IPC Business Press Ltd. ©.]

liquid is more readily absorbed, thereby leading to enhanced plasticization and a lowering of T_g. In turn, this would provide for greater chain mobility and allow crazing to occur at a lower stress level.

Miltz *et al.* [55] also examined the effect of a liquid environment's solubility parameter on the static and fatigue properties of polycarbonate. They found that when $|\delta_S - \delta_P|$ approached zero, the craze growth rate at the root of a premachined notch decreased, whereas dissolution and cracking became more extensive. Also, they noted that the effect of liquid environments on the fatigue and static fracture life of PC depended in part on whether a crack existed prior to the time of exposure to the hostile environment.

Finally, a study of the effect of atmospheric conditions in Sapporo, Japan, on the fatigue strength of PS, PVC, PMMA, ABS, and polyoxymethylene (POM) was conducted by Suzuki and Tsurue [56]. After one and two years exposure to natural elements, they measured tensile, bending, and fatigue strength ($N = 10^7$ cycles) and compared these values with nonexposed reference test bars. The fatigue strengths of weathered PS, ABS, and POM decreased markedly both in an absolute sense and relative to the samples' tensile and bending strengths. By contrast, the mechanical response of PVC and PMMA samples was not altered significantly after weathering. This difference in weathering response was related consistently to the quality of the exposed specimen surface. Fine cracks were found in PS and ABS with POM exhibiting a chalked appearance. On the other hand, the surfaces of the PVC and PMMA samples changed little. These results demonstrate that the fatigue initiation stage of cyclic damage is truncated by weathering, which leads to an overall reduction in fatigue life. Since Warty *et al.* [48] found that the fatigue life of polystyrene was not adversely affected by water, the deterioration noted by Suzuki and Tsurue [56] in the polystyrene samples must have been attributed to other environmental factors than the amount of precipitation. A further discussion of environmental effect on fatigue crack propagation in polymeric solids is found in Section 3.6.

Even though numerous studies have been conducted in the past to identify relationships between various material properties and molecular weight M, little work has been done to explore fatigue property dependence on M. Of particular note, Sauer and co-workers [57, 58] found more than a two-decade improvement in the fatigue life in polystyrene when molecular weight was increased from 1.6×10^5 to 2×10^6 (Fig. 2.23). This improvement in fatigue life with increasing M is somewhat surprising since one finds only small changes in such basic properties as elastic modulus, yield strength, and glass transition temperature when M is altered by this amount. A similar improvement in fatigue performance was identified in crystalline poly-

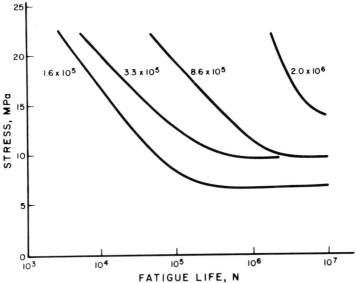

Fig. 2.23 Effect of molecular weight on fatigue life in polystyrene [Sauer, Foden, and Morrow (58)].

ethylene [58]. Additional studies by Sauer and co-workers [59, 60] have focused on attempts to determine whether polymer fatigue properties were more sensitive to the number-average molecular weight $\overline{M}_n$ as opposed to the weight-average molecular weight $\overline{M}_w$. Specimens were irradiated with a Co^{60} γ source for various times, which led to chain branching by a cross-linking reaction. In this manner, $\overline{M}_n$ could be maintained relatively constant while $\overline{M}_w$ and MWD increased. Fatigue test results showed that fatigue life changed little with increasing $\overline{M}_w$ when $\overline{M}_n$ remained constant. Recall that other results by Sauer indicated a large improvement in fatigue life with increasing molecular weight. When other samples with varying $\overline{M}_n$ were tested, a good correlation was found between fatigue life and $\overline{M}_n$. Based on these findings, Sauer *et al.* [60] concluded that fatigue life was more strongly dependent on $\overline{M}_n$ than $\overline{M}_w$. Further studies are needed to confirm this observation and to clarify certain points. For example, the irradiation treatment used to increase $\overline{M}_w$ while maintaining $\overline{M}_n$ at a constant level, converts the linear polystyrene polymer to a crosslinked solid. It is not known what effect this will have on deformation and fatigue behavior, though some preliminary fatigue crack propagation studies show cross-linking to be detrimental to fatigue resistance (see Section 3.8.3). Therefore, any beneficial effects due to a larger $\overline{M}_w$ may be offset by enhanced cross-linking. It might be best to examine fatigue life at different $\overline{M}_w$ and $\overline{M}_n$ combinations while maintaining the material in the uncrosslinked condition. Indeed, Sauer *et al.* reported "... that for the same $\overline{M}_n$ value, the polymer having the broad molecular weight distribution ($\overline{M}_w/\overline{M}_n = 2.74$) exhibits a greater resistance to [fatigue] fracture than the polymers of narrow MWD" [60]. In an overall sense, the improvement in fatigue performance with increasing molecular weight is believed due to greater entanglement density and a propensity for greater orientation hardening of microfibrils that span the crack tip crazes. Both factors should stabilize the craze and prolong its cyclic life. The effect of molecular weight on fatigue crack propagation behavior and fracture surface micromorphology is discussed in Sections 3.8.2 and 4.2.2, respectively.

The effect of mean stress on fatigue life of polystyrene has been examined by Sauer *et al.* [61]. By conducting tests under constant stress amplitude at varying mean stress and constant maximum stress with varying mean and alternating stress test conditions, they found that the fatigue life of polystyrene decreased sharply with increasing stress range at a constant mean stress level and also decreased at a fixed stress range with increasing mean stress. A similar finding regarding mean stress on fatigue crack propagation in precracked polystyrene test plates is discussed in Section 3.7. As a final note, Sauer *et al.* [61] found that their specimens exhibited almost an order of magnitude greater fatigue life than that reported by Rabinowitz *et al.* [33]

for the same mean and alternating stress conditions. Sauer *et al.* [61] attributed this difference to a larger $\overline{M}_w$ in their samples and to their use of a higher test frequency (26 versus 0.1 Hz), which would be expected to lessen any creep-enhanced damage during the fatigue test.

A preliminary investigation of the effect of loading history on the fatigue life of polycarbonate has been reported recently by Mackey *et al.* [62]. They subjected fatigue bars to a preliminary tensile stress 90–95% of the anticipated yield strength before conducting some fatigue tests. They found that the fatigue life in cyclic load control was greater at lower stress amplitude levels when the samples were prestrained to introduce crazes. These authors argued that the precrazing procedure was analogous to a beneficial addition of a soft second phase. They did not, however, speculate as to why the fatigue life of prestrained samples was reduced at high alternating stresses relative to the unstrained specimens, Certainly more studies like this are needed to clarify the impact of prior stress and strain history on total component fatigue life. A brief discussion of this subject as it pertains to fatigue crack propagation is found in Section 3.7.

2.7 EVALUATION OF A FATIGUE FRACTURE RATE THEORY

To date, analysis of polymer fatigue behavior has been limited primarily to phenomenological evaluations of fatigue life as a function of either cyclic stresses or strains. With one exception, almost no attention has been given toward establishing a fatigue fracture model. During the past decade, numerous Russian authors have examined the possibility that fatigue fracture could be evaluated in the same manner as creep fracture, i.e., in terms of a reaction rate theory [63–69]. It has been argued that the fracture process involves cumulative primary bond rupture enhanced by thermal fluctuations that can be expressed by a relationship of the form

$$\tau = \tau_0 \exp[(U_0 - \gamma\sigma)/kT], \tag{2.9}$$

where τ is the component or specimen life, τ_0 the material constant associated with the period of natural vibration of atoms within the solid, σ the applied stress, k the Boltzmann's constant, U_0 the activation energy of the fracture process (primary bond rupture), T the absolute temperature and, γ the structure sensitive material property.

While creep data have been analyzed somewhat effectively with Eq. (2.9), the use of this relationship for the case of cyclic loading has not met with considerable success. For example, one finds that creep life τ_{creep} of a specimen is often much greater than its corresponding fatigue life τ_{fatigue}. Part of this discrepancy has been attributed to cyclic stress-induced hysteretic heating that raises the temperature of the specimen and causes a sizable increase in the reaction rate [64–66]. It has been argued that the remaining difference between τ_{creep} and τ_{fatigue} can be traced to insufficient time for stress relaxation under cyclic loading conditions; stress relaxation was thought to be necessary to equalize local stress concentrations at newly formed defects within the solid. In support of this hypothesis, it was demonstrated that the ratio $\tau_{\text{creep}}/\tau_{\text{fatigue}}$ approached unity when the fatigue test temperature was increased and/or the test frequency decreased. It should be recognized that such test conditions would generate similar damage mechanisms in creep and fatigue, thereby leading to a uniform activation energy. Agreement between fatigue and creep lives would then be expected.

Since it was originally assumed that U_0 remained constant and not sensitive to the details of loading or material structure, the observed differences between τ_{creep} and τ_{fatigue} had been related to changes in the structure sensitive property γ [64–66]. However, Regel and Lekosovskii after reviewing the available data concluded that these data "... do not give grounds for categorically rejecting other possible ways of accounting for the fatigue effect, including changes in the activation energy of the fracture process" [66]. Indeed, Weaver and Beatty [70] have shown that the activation energy for compressive fatigue in polystyrene is dependent on temperature (Fig. 2.24).

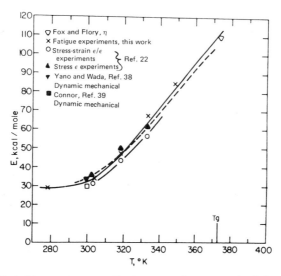

Fig. 2.24 Effect of temperature on activation energy for compressive fatigue in polystyrene. Data symbols refer to cited results from original source [Weaver and Beatty (70)].

Furthermore, the inferred values of the activation energy were interpreted as reflecting processes associated with localized plastic deformation as opposed to the energy for primary bond rupture as originally assumed by Zhurkov. This latter approach seems more reasonable since it would not be expected that cyclic disentanglement processes at ambient temperatures associated with fatigue loading would involve the same activation energy as that associated with monotonic loading at elevated temperatures.

It must be concluded from this discussion that the fatigue fracture process is complex and involves a cooperative or even synergistic interaction among several different deformation mechanisms, each with its own characteristic activation energy and kinetics.

2.8 USE OF CYCLIC STRESS AND STRAIN DATA IN COMPONENT DESIGN

Until recently, few examples could be found for the application of polymeric solids in engineering components that would be subjected to significant cyclic loading. Consequently, polymer component design procedures for such usage can be discussed only in general terms.

It is clear from the preceding discussion that certain laboratory fatigue results may or may not be applicable in the engineering design of a particular polymeric component. This is due to the fact that there are differences in test versus service cyclic frequency, specimen surface area-to-volume ratio, cooling conditions, waveform, and nature of loading (fixed loads or fixed strains), which preclude the use of certain data in certain instances. Accordingly, numerous authors have repeatedly cautioned their readers to conduct laboratory tests that most closely correspond to the service conditions of the components they wish to design.

Perhaps the greatest single factor to be established is whether the component will experience cyclic loadings that are predominantly load-controlled as opposed to deflection-controlled. This has a strong bearing on whether or not thermal fatigue would occur. For example, it is known that early life failure due to thermal melting can occur in polymeric gears under load-controlled conditions [71–73]. A significant heat buildup along the pitch line of gear teeth can occur as a result of rubbing and internal heat generation. For this reason, gears are often operated in the presence of a lubricant that lowers tooth flank temperature by reducing surface friction and providing a media for more effective heat transfer. Tooth temperature can also be reduced by lowering the gear stress and the gear speed. Figure 2.25 reveals the marked improvement in fatigue life when the tooth temperature is

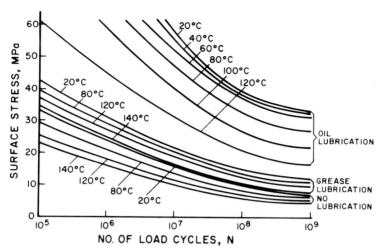

Fig. 2.25 Improved gear fatigue life resulting from lubrication and lower tooth temperature [Hackman and Strickle (71)].

decreased and the gears are lubricated. As was stated earlier in Section 2.3.4, the avoidance of thermal failure does not preclude the possibility for mechanical fatigue fracture. In gears, fatigue cracks can be nucleated at the tooth root and grow to fracture. The tooth root is a highly likely site for such cracking since the tooth bending stresses are a maximum at this location.

Since so little is known about the phenomenology of cyclic strain testing in polymers, a discussion of polymer component design under this loading condition is somewhat premature. Instead, the reader is referred to procedures established for various metallic solids [40, 41, 74]. *(but these can not actually be used with plastics !!)*

REFERENCES

[1] Military Specification MIL-A-83444, USAF (2 July 1974).

[1a] M. N. Riddell, *Plast. Eng.* **30**(4), 71 (1974).

[2] ASTM Standard D671-71, Part 27, p. 216 (1973).

[3] B. J. Lazan, *ASME Trans.* **65**, 87 (1943).

[4] B. J. Lazan and A. Yorgiadis, ASTM STP 59, p. 66 (1944).

[5] W. N. Findley, *Proc. ASTM* **41**, p. 1231 (1941).

[6] S. B. Ratner and N. I. Barash, *Mekh. Polim.* **1**(1), p. 124 (1965).

[7] S. B. Ratner and V. I. Korobov, *Mekh. Polim.* **1**(3), 93 (1965).

[8] P. P. Oldyrev, *Mekh. Polim.* **3**(1), 111 (1967).

[9] S. B. Ratner and A. V. Stinskas, *Mekh. Polim.* **3**(1), 179 (1967).

[10] P. P. Oldyrev and U. P. Tamuzh, *Mekh. Polim.* **3**(5), 864 (1967).

[11] P. P. Oldyrev, *Mekh. Polim.* **3**(3), 483 (1967).

[12] S. B. Ratner and S. T. Buglo, *Mekh. Polim.* **5**(3), 465 (1969).

[13] M. N. Riddell, G. P. Koo, and J. L. O'Toole, *Polym. Eng. Sci.* **6**, 363 (1966).
[14] G. P. Koo, M. N. Riddell and J. L. O'Toole, *Polym. Eng. Sci.* **7**, 182 (1967).
[15] L. C. Cessna, J. A. Levens, and J. B. Thomson, *Polym. Eng. Sci.* **9**(5), p. 339 (1969).
[16] A. X. Schmidt and C. A. Marlies, "Principles of High-Polymer Theory and Practice," p. 578. McGraw-Hill, New York, 1948.
[17] J. D. Ferry, "Viscoelastic Properties of Polymers." Wiley, New York, 1961.
[18] R. J. Crawford and P. P. Benham, Polymer **16**, p. 908 (1975).
[19] L. J. Broutman and S. K. Gaggar, *Int. J. Poly. Mater.* **1**, 295 (1972).
[20] A. V. Stinskas and S. B. Ratner, *Sov. Plast.* **49** (1962).
[21] K. Oberbach, *Kunstoffe* **55**, 356 (1965).
[22] D. C. Prevorsek, G. E. Lamb, and M. L. Brooks, *Polym. Eng. Sci.* **7**, 269 (1967).
[23] D. A. Opp, D. W. Skinner, and R. J. Wiktorek, *Polym. Eng. Sci.* **9**, 121 (1969).
[24] C. E. Feltner and J. D. Morrow, *J. Bas. Eng. Trans. ASME Ser. D* **83**, 15 (1961).
[25] K. V. Gotham, *Plast. Polym.* **41**(156), 273 (1973).
[26] E. H. Andrews, "Testing of Polymers" (W. Brown, ed.), Vol. IV, p. 237. Wiley (Interscience), New York, 1969.
[27] I. Constable, J. G. Williams, and D. J. Burns, *J. Mech. Eng. Sci.* **12**, 20 (1970).
[28] R. Attermo and G. Ostberg, *Int. J. Fracture Mech.* **7**, 122 (1971).
[29] G. I. Barenblatt, V. M. Entov, and R. L. Salganik, *IUTAM Symp. Thermoinelast., East Kilbridge, June 25* (1968).
[30] ASTM Book of Standards, Part 27, p. 916 (1973).
[31] R. Gauvin and J. P. Trotignon, *J. Test. Eval.* **6**(1)1, p. 48 (1978).
[32] S. S. Manson and M. H. Hirschberg, Fatigue, An Interdisciplinary Approach, *Proc. Sagamore Army Res. Conf., 10th* (J. J. Burke and V. Weiss, eds.), p. 133. Syracuse Univ. Press, New York, 1964.
[33] S. Rabinowitz, A. R. Krause, and P. Beardmore, *J. Mater. Sci.* **8**, p. 11 (1973).
[34] S. Rabinowitz and P. Beardmore, *J. Mater. Sci.* **9**, 81 (1974).
[35] P. Beardmore and S. Rabinowitz, "Polymeric Materials," p. 551. American Society of Metals, Metals Park, Ohio, 1975.
[36] P. Beardmore and S. Rabinowitz, "Treatise on Materials Science and Technology" (R. J. Arsenault, ed.), Vol. 6, p. 267. Academic Press, New York, 1975.
[37] B. Tomkins and W. D. Biggs, *J. Mater. Sci.* **4**, 532 (1969).
[38] P. Beardmore and S. Rabinowitz, "Polymeric Materials," p. 539. American Society of Metals, Metals Park, Ohio, 1975.
[39] P. Beardmore and S. Rabinowitz, "Polymeric Materials," p. 544. American Society of Metals, Metals Park, Ohio, 1975.
[40] R. W. Landgraf, ASTM STP 467, p. 3 (1970).
[41] R. W. Landgraf, J. D. Morrow, and T. Endo, *J. Mater.* **4**(1), 176 (1969).
[42] L. F. Coffin, Jr., *Trans. ASME* **76**, 931 (1954).
[43] S. S. Manson, Rep. 1170, NACA (1954).
[44] I. P. Bareishis and A. V. Stinskas, *Mekh. Polim.* **9**(3), 562 (1973).
[45] T. I. Sogolova, *Mekh. Polim.* **2**(5), 643 (1966).
[46] V. G. Savkin, V. A. Belyi, T. I. Sogolova, and V. A. Kargin, *Mekh. Polim.*, **2**(6), 803 (1966).
[47] V. K. Kuchinskas and A. N. Machyulis, *Mekh. Polim.* **3**(4), 713 (1967).
[48] S. Warty, D. R. Morrow, and J. A. Sauer, *Polymer*, to be published.
[49] S. G. Agamalyan, *Mekh. Polimerov* **10**(3), 107 (1974).
[50] A. V. Stinskas, Y. P. Baushis, and I. P. Bareishis, *Mekh. Polim.* **8**(1), 59 (1972).
[51] Y. N. VanGaut, *Mekh. Polim.* **1**(3), 151 (1965).
[52] A. N. Machyulis, M. I. Pugina, A. A. Zhechyus, V. K. Kuchinskas, and A. P. Stasyunas, *Mekh. Polim.* **2**(1), p. 60 (1966).

[53] I. P. Bareishis and A. V. Stinskas, *Mekh. Polim.* **8**(2), 875 (1972).

[54] L. Lee, *Adv. Chem. Ser.* **87,** 106 (1968).

[55] J. Miltz, A. T. DiBenedetto, and S. Petrie, *J. Mater. Sci.* **13,** 2037 (1978).

[56] S. Suzuki and T. Tsurue, *Int. Conf. Mech. Beh. Mater.*, *2nd, Boston, Massachusetts.* American Society of Metals, Metals Park, Ohio, 1976.

[57] E. Foden, D. R. Morrow, and J. A. Sauer, *J. Appl. Polym. Sci.* **16,** 519 (1972).

[58] J. A. Sauer, E. Foden, and D. R. Morrow, *Polym. Eng. Sci.* **17,** p. 246 (1977).

[59] J. A. Sauer, *Polymer* **19,** 859 (1978).

[60] S. Warty, J. A. Sauer, and A. Charlesby, *Eur. Polym. J.* **15**(5), 445 (1979).

[61] J. A. Sauer, A. D. McMaster, and D. R. Morrow, *J. Macromol. Sci.-Phys.* **B12**(4), 535 (1976).

[62] M. E. Mackay, T. G. Teng, and J. M. Schultz, *J. Mater. Sci.* **14,** 221 (1979).

[63] G. M. Bartenev, *Mekh. Polim.* **2**(5), 700 (1966).

[64] O. F. Kireenko, A. M. Leksovskii, and V. R. Regel, *Mekh. Polim.* **4**(3), 483 (1968).

[65] A. M. Leksovskii and V. R. Regel, *Mekh. Polim.* **4**(4), 648 (1968).

[66] V. R. Regel and A. M. Leksovskii, *Mekh. Polim.* **5**(1), 70 (1969).

[67] A. M. Leksovskii and V. R. Regel, *Mekh. Polim.* **6**(2), 253 (1970).

[68] S. N. Zhurkov and V. S. Kuksenko, *Mekh. Polim.* **10**(5), 792 (1974).

[69] A. M. Leksovskii and B. Gaffarov, *Mekh. Polim.* **12**(5), 912 (1976).

[70] J. L. Weaver and C. L. Beatty, *Polym. Eng. Sci.* **18**(14), 1117 (1978).

[71] H. Hachmann and E. Strickle, *Soc. Plast. Eng.*, *Ann. Tech. Conf.* **26,** 512 (1968).

[72] S. S. Yousef and D. J. Burns, *Mech. Machine Theory* **8,** 175 (1973).

[73] H. Yelle and M. Poupard, *Polym. Eng. Sci.* **15**(2), 90 (1975).

[74] B. I. Sandor, "Fundamentals of Cyclic Stress and Strain." Univ. Wisconsin Press, Madison, Wisconsin, 1972.

3 | *Fatigue Crack Propagation*

3.1 INTRODUCTION

Attention will now be given to a study of the fatigue life of laboratory samples and engineering components that contain preexistent defects. In this manner, one may ignore the number of loading cycles needed for crack initiation and, rather, consider total fatigue life as being dominated by the kinetics of crack propagation. In addition to focusing attention on a single aspect of fatigue damage accumulation, testing of a precracked test bar as opposed to an unnotched specimen represents a realistic step toward laboratory test simulation of actual service conditions. Surely many engineering components contain numerous adventitious flaws capable of growth under applied cyclic loads or displacements. Examples of such defects would include voids, sprue marks, flash lines, welds, foreign particles, overall surface finish, and various machine marks. Consequently, component fatigue life may well be dominated by crack propagation processes rather than by initiation. The fracture surface of a plastic chair leg shown in Fig. 3.1 serves as a prime example of the growth of a fatigue crack from an extensive array of internal voids that resulted from improper molding procedure(s). The series of concentric rings emanating from the porous interior region represent successive periods of growth of the fatigue crack and are referred to in the fatigue literature as "clam shell" or "beach" markings [1]. Note that similar concentric rings were found on the fracture surface of the plastic tricycle shown in Fig. 2.1.

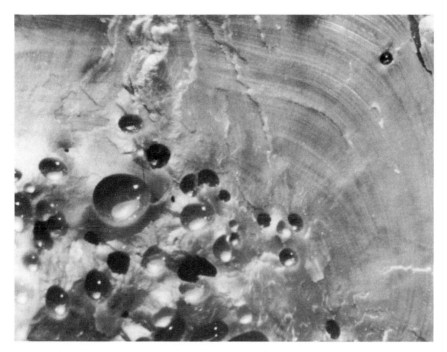

Fig. 3.1 Fracture surface of plastic chair leg. Note fatigue rings emanating from array of internal voids.

The task, then, is to develop a meaningful laboratory test that provides a maximum amount of information about the cracking stage of fatigue damage. The most straightforward procedure might appear to involve the generation of S–N plots from samples containing deliberately introduced defects. However, one would have to record the number of loading cycles needed to break the sample at any given stress level for a wide variety of notch geometries. Furthermore, one would still not know with any degree of certainty how much cyclic life would remain after the initial defect had grown to some larger dimension. Recognizing this limitation, engineers and researchers have sought to characterize the *kinetics* of the crack growth process by monitoring the rate of crack advance as a function of the instantaneous crack length and applied stress level. This procedure requires that a technician (or computer) devote constant attention to the cracking process in a given sample, but yields an enormous increase in test data as compared with the recording of only the total cycles to failure.

A very large body of literature concerned with the kinetics of fatigue crack growth has been amassed over the past two decades. Much of this information has dealt with an analysis of fatigue cracking rates in terms of fracture mechanics parameters. Since many readers may not be familiar with this literature and with the fundamental concepts of fracture mechanics, it is appropriate at this point to deal with these topics. Following this, the reader should be better equipped to evaluate the specific topics to be covered later in this and subsequent chapters.

3.2 ELEMENTS OF FRACTURE MECHANICS CONCEPTS

A major objective of any fracture theory is to identify a quantitative relationship between the stresses acting on a structure, the dimensions, and configuration of a crack existing in that structure and the relevant mechanical properties of the material in question. Griffith [2] initially considered this problem almost 60 yr ago for the case of an ideally brittle solid. His analysis was based on a thermodynamic energy rate model involving the incremental advance of a crack Δa driven by the strain energy release rate $\partial u/\partial a$, and an energy sink that would absorb the input energy γ_s (the latter representing the surface energy of the material). Instability of the crack was believed to occur when the strain energy release rate associated with an incremental advance of the crack was greater than the energy necessary to create the free surfaces of the newly generated crack increment. Using the strain energy solution developed by Inglis [3] for the case of a center-cracked panel, Griffith showed that the crack would grow unstably when

$$\sigma = (2E\gamma_s/\pi a)^{1/2}, \tag{3.1}$$

where σ is the gross applied stress, E the modulus of elasticity, γ_s the surface energy of the solid, and a half the crack length of the center-notched panel. Fracture experiments performed on hollow, cracked rods and spheres of glass provided a measure of surface energy [see Eq. (3.1)], which was in very good agreement with an extrapolated value of surface energy based on surface tension experiments. Later, Orowan [4] suggested that Eq. (3.1) could be extended to the case for metals with the addition of another energy sink term, the energy associated with plastic deformation. This modification then gives

$$\sigma = [2E(\gamma_s + \gamma_p)/\pi a]^{1/2}, \tag{3.2}$$

where γ_p is the plastic deformation energy term and $\gamma_p \gg \gamma_s$. Therefore, the total fracture energy γ_f (i.e., $\gamma_s + \gamma_p$) would be found to be much larger than γ_s. This fact has been demonstrated for the case of polymer fracture where γ_p was shown to be a function of strain rate [5–8]. (Recall Section 1.4.1.)

Irwin [9] also considered the fracture of solids process from a thermodynamic point of view and arrived at a relationship similar to that expressed by Griffith

$$\sigma = (E\mathscr{G}/\pi a)^{1/2},$$ (3.3)

where $\mathscr{G}$ is the strain energy release rate $\partial u/\partial a$. Note that the stress–flaw size relationship is now defined in terms of a strain energy release rate $\mathscr{G}$ instead of the energy sink terms γ_s and γ_p. Comparing Eqs. (3.2) and (3.3), it is seen that $\mathscr{G} = 2(\gamma_s + \gamma_p)$.

The fracture of flawed components also may be analyzed by a stress analysis involving concepts of elastic theory. Based on the method of Westergaard [10], Irwin was able to describe the stress field at a crack tip using the notation shown in Fig. 3.2 such that

$$\sigma_y = \frac{K}{(2\pi r)^{1/2}} \cos\frac{\theta}{2}\left[1 + \sin\frac{\theta}{2}\sin\frac{3\theta}{2}\right],$$

$$\sigma_x = \frac{K}{(2\pi r)^{1/2}} \cos\frac{\theta}{2}\left[1 - \sin\frac{\theta}{2}\sin\frac{3\theta}{2}\right],$$

$$\tau_{xy} = \frac{K}{(2\pi r)^{1/2}} \cos\frac{\theta}{2}\cos\frac{3\theta}{2}\sin\frac{\theta}{2},$$ (3.4)

$$\tau_{xz} = \tau_{yz} = 0,$$

$$\sigma_z \approx v(\sigma_y + \sigma_x) \quad \text{(plane strain)},$$

$$\sigma_z \approx 0 \quad \text{(plane stress)},$$

where v is Poisson's ratio and K the stress intensity factor—a function of σ and a.

It is apparent from Eq. (3.4) that these local stresses must rise to extremely high levels as r approaches zero. However, this circumstance is precluded by the onset of plastic deformation at the crack tip. The extent of the yielded zone around the crack tip in plane stress has been estimated [11] from Eq. (3.4) by setting the local stress equal to the yield strength so that

$$r_y \approx \frac{1}{2\pi}\frac{K^2}{\sigma_{ys}^2},$$ (3.5)

where σ_{ys} is the material yield strength and r_y the crack-tip plastic zone radius.

For the case of plane strain where plastic constraint restricts the development of the plastic zone, the size is estimated [12] to be

$$r_y \approx \frac{1}{6\pi}\frac{K^2}{\sigma_{ys}^2}.$$ (3.6)

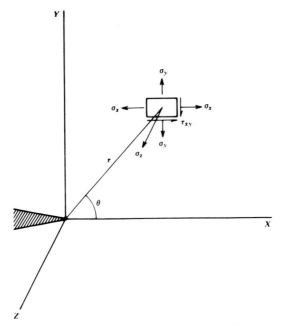

Fig. 3.2   Stress distribution around crack tip.

For additional comments concerning the crack tip plastic zone see Section 3.8.1. Whether plane strain or plane stress conditions will apply in a given situation depends upon the size of the plastic zone relative to the component thickness and actual crack length. For example, plane strain fracture conditions are considered to be present when the plastic zone size [computed from Eq. (3.6)] is equal to or less than 2% of both the component thickness and crack length [13].

An important feature of Eq. (3.4) is the fact that the stress distribution around any crack in a structure is similar and depends only on the parameters r and θ. The difference between one cracked component and another lies in the magnitude of the stress intensity factor K; i.e., K serves as a scale factor to define the magnitude of the crack tip stress field. It has been shown that $K = f(\sigma, a)$ where the functionality depends on the configuration of the cracked component and the manner in which the loads are applied. Many functions have been determined for various specimen configurations and are available from the fracture mechanics literature [14–16]. These solutions include those for certain specimen configurations that are used extensively in fatigue crack propagation testing, such as the single edge notched (SEN) and compact tension (CT) specimens shown in Fig. 3.3.

In the simplest case—that of an infinitely large panel with a small central

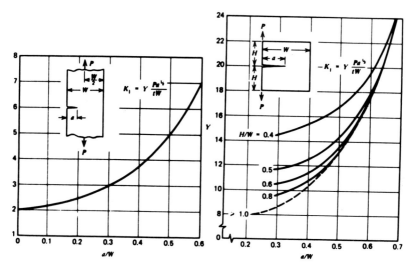

Fig. 3.3 Stress intensity factor solutions for (a) single edge-notched and (b) compact tensile specimens. [R. W. Hertzberg, "Deformation and Fracture Mechanics of Engineering Materials." © 1976 by John Wiley & Sons, Inc. Reprinted by permission of John Wiley & Sons, Inc.]

crack of length $2a$—the stress intensity factor is given by

$$K = \sigma(\pi a)^{1/2}. \tag{3.7}$$

By combining Eqs. (3.3) and (3.7), it is seen that $\mathscr{G}E = K^2$ (plane stress) and $\mathscr{G}E \cdot (1 - v^2)^{1/2} = K^2$ (plane strain), representing relationships that were established on a more rigorous basis by Bueckner [17]. Equation (3.8) represents the more general case

$$K = Y\sigma a^{1/2}, \tag{3.8}$$

where Y is the geometrical factor $f(a/w)$ and W the specimen width. The Y calibration for the SEN and CT type specimens is shown in graphical form in Fig. 3.3 where $\sigma = P/tW$ (t = thickness). Once the stress intensity factor for a given test sample is known, it is then possible to determine the maximum stress intensity factor that would cause failure. This critical value K_c is described in the literature as the *fracture toughness* of the material. Under plane strain test conditions, the fracture toughness value takes on a special designation, K_{Ic}. Plane strain fracture toughness values are determined by adherence to specific test procedures noted in ASTM Specification E399-78 [13]. Fatigue crack propagation would then proceed in a stable manner across the width of a component or test specimen until the magnitude of the crack-tip stress intensity factor became equal to K_c or K_{Ic}. Rapid unstable fracture would then occur. It should be noted that the final stress intensity level in a fatigue test may not correspond exactly to K_c or K_{Ic} values obtained in accordance with ASTM E399-78 requirements. This arises from the fact that test conditions are different for the fatigue and fracture toughness tests. Nevertheless, the final K level attained in the fatigue test can serve as a good indicator of the material's fracture resistance; as such, it represents an "apparent" fracture toughness.

3.3 ELEMENTS OF FATIGUE CRACK PROPAGATION (FCP)

As was mentioned in Section 3.1, the kinetics of the fatigue crack propagation (FCP) process can be examined by simply measuring the change in crack length of a precracked sample as a function of the total number of loading cycles. Many monitoring techniques have been employed such as compliance measurements, acoustic emission detectors, eddy current techniques, electropotential measurements, and the use of a calibrated traveling microscope. Most of the FCP data generated to date for polymeric solids have been based on traveling microscope readings. A typical plot of such data is shown in Fig. 3.4, which shows the crack length increasing with

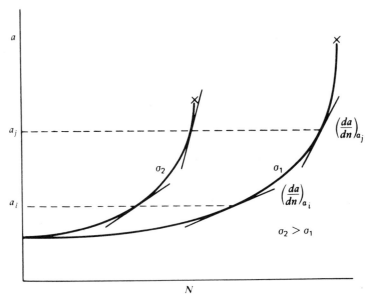

Fig. 3.4 Crack length versus number of loading cycles. Fatigue crack growth rate increases with increasing stress level and crack length. [R. W. Hertzberg, "Deformation and Fracture Mechanics of Engineering Materials," © 1976 by John Wiley & Sons, Inc. Reprinted by permission of John Wiley & Sons, Inc.]

number of loading cycles. The fatigue crack growth rate per cycle da/dN is determined from such a curve at any value of crack length by graphical procedures or by computation. For most specimen configurations, the crack growth rate increases with increasing crack length, thereby shortening component life at an alarming rate. From this, it is recognized that most of the loading cycles involved in the total life of an engineering component are consumed during the early stages of crack extension when the crack is small and, possibly, undetected.

From Fig. 3.4, it is also seen that da/dN increases with increasing stress level such that

$$da/dN \propto f(\sigma, a). \tag{3.9}$$

Numerous relationships have been proposed to describe the FCP behavior of metallic and polymeric solids, based on empirical formulations and fracture mechanics principles. The latter approach has proven to be most flexible and has become widely adopted. Paris [18–20] postulated that the stress intensity factor, itself a function of stress and crack length, was the major controlling factor in the FCP process. This suggestion is totally consistent with the fact that the stress intensity factor controls static fracture and environment-assisted cracking as well [1]. From Fig. 3.4 then, the growth rate $(da/dN)a_{i,j,\ldots,n}$ at any arbitrary crack length $a_{i,j,\ldots,n}$ would correspond to respective values of $K_{i,j,\ldots,n}$ (for some fixed stress level). Paris found that the key stress variable was the stress range $(\sigma_{max} - \sigma_{min})$ and so described da/dN values in terms of the stress intensity factor range ΔK with a relationship of the form

$$da/dN = A\Delta K^m, \tag{3.10}$$

where da/dN is the fatigue crack growth rate, ΔK the stress intensity factor range $(\Delta K = K_{max} - K_{min})$, and $A, m = f$ (material variables, environment, frequency, temperature, stress ratio). Though many studies have verified the log–log linear relationship between da/dN and ΔK such as that shown in Fig. 3.5 for several polymeric solids, others have shown FCP plots to assume a sigmoidal shape; crack growth rates are sometimes found to decrease to vanishingly low values as ΔK approaches some limiting threshold value ΔK_{th} and increase to very high values as K_{max} approaches K_c. The dependence of fatigue crack propagation in polymers and metals on ΔK has been reviewed previously and may be referred to by the reader [1, 21].

The earliest applications of fracture mechanics to FCP in polymers made use of the energy rate approach. By modifying the Griffith equation, Rivlin and Thomas [22] were able to describe the crack instability condition in a more general sense by not requiring the material to obey classical theory of elasticity. Several solutions were developed [23] to quantify the strain energy

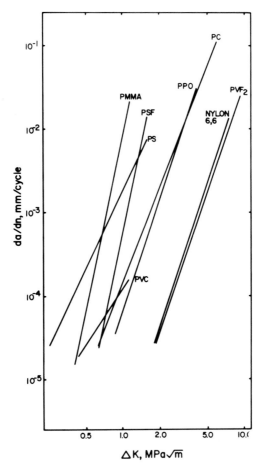

Fig. 3.5 Fatigue crack propagation behavior in selected crystalline and amorphous polymers [Hertzberg, Manson, and Skibo (34)].

release rate for different specimen geometries. For the case of a SEN panel, e.g.,

$$\mathcal{T} = k_1 2aU, \tag{3.11}$$

where $\mathcal{T}$ is the surface work term, related to $\gamma_s + \gamma_p$ [see Eq. (3.2)], k_1 a constant that varies slowly with the material, a the crack length, and U the uniform strain energy density. For a completely elastic uniformly stressed material, $U = \sigma^2/2E$ and with $k_1 = \pi$, it is seen that $\mathcal{T}$ in Eq. (3.11) is equivalent to $\mathcal{G}$ [Eq. (3.3)]. When the stress–strain relationship of a material is nonlinear, U must be determined separately [23].

The Rivlin–Thomas approach was used initially to examine the fatigue response of rubbers [23–28]. Thomas [28] found that at least over a reasonable range the fatigue crack growth rate range in natural rubber could be expressed in terms of the energy parameter $\mathscr{T}$ by an equation of the form

$$da/dN = A\Delta\mathscr{T}^n \qquad (3.12)$$

where da/dN is the crack growth rate per cycle, A the material constant, $\Delta\mathscr{T}$ the surface work parameter range, and n the exponent, approximately equal to 2. Note the strong similarity between Eqs. (3.10) and (3.12).

Attention is now given to detailed studies of the effects of external and internal variables on the FCP behavior in polymeric solids. To begin, crack growth rates will be examined as a function of such experimental test variables as frequency, test temperature, environment, and mean stress level. Then attention is given to the effect of molecular properties and composition on fatigue crack propagation. Finally, there is a discussion dealing with design philosophy and design life calculations based on fatigue crack propagation information.

3.4   EFFECT OF TEST FREQUENCY ON FCP

In Chapter 2, a clear distinction was made between thermal and mechanical fatigue failure modes in polymeric solids. It was shown that thermal failures occurred by adiabatic heating that produced an increase in specimen temperature; the extent of such heat buildup was found to vary directly with cyclic frequency. It has been argued that unstable heat buildup is encountered in situations where the volume of material that generates the heat (e.g., the test specimen gauge section) exceeds some critical quantity for a given set of test conditions. On the other hand, if the zone of highly stressed material were to be localized, as at the crack tip of a prenotched sample, the material surrounding the crack-tip damage zone and the surrounding environment might be capable of conducting away enough heat so as to preclude an unstable temperature rise. Consequently, it is of interest to examine how the fatigue crack propagation resistance of engineering polymers varies as a function of test frequency.

Indeed, the results of such studies are most striking. When prenotched samples were tested over a range of cyclic frequencies from 0.1 to 100 Hz, the associated FCP rates for several polymers such as poly(methyl methacrylate) (PMMA) [29–33], polystyrene (PS) [34, 35], poly(vinyl chloride) (PVC) [36–39], and a poly(phenylene oxide) (PPO)/high-impact polystyrene (HIPS) blend [34] *decreased* with increasing frequency. Other polymers such as

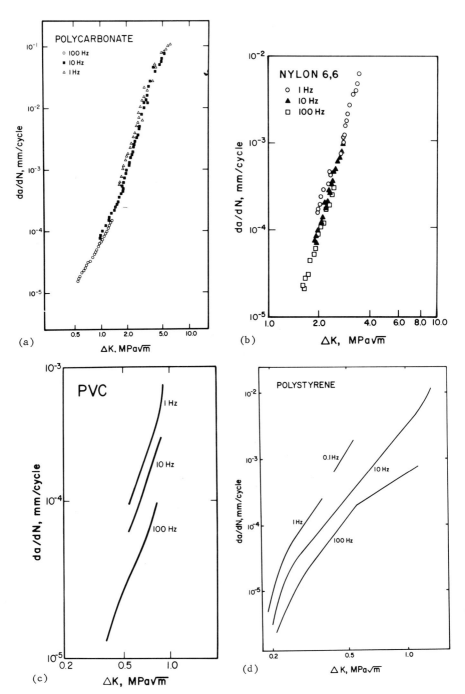

Fig. 3.6 Effect of cyclic frequency on FCP rates in (a) polycarbonate [Hertzberg, Manson, and Skibo (34)]; (b) nylon 66 [R. W. Hertzberg, M. D. Skibo, J. A. Manson, and J. K. Donald, *J. Mater. Sci.* **14**, 1979, 1754, Chapman and Hall, Ltd.]; (c) poly(vinyl chloride) [Skibo (55)]; (d) polystyrene [Hertzberg, Manson, Skibo (34)].

polycarbonate (PC) [21, 40], polysulfone (PSF) [21], nylon 66 [41], and
poly(vinylidene fluoride) (PVDF) [34] showed no apparent sensitivity of
fatigue crack propagation rate to test frequency. A sampling of such data is
shown in Fig. 3.6. Conflicting FCP results have been reported for nylon 66 by
Arad *et al.* [42, 43] and El-Hakeem [44] who reported that FCP rates decreased with increasing test frequency. The reason for this difference in
behavior with that shown in Fig. 3.6b is not clear at this time, though the
much higher mean stress used in the Arad *et al.* experiments may have been a
contributing factor. The frequency sensitivity factor (FSF), defined as the
multiple by which the FCP rate changes per decade change in test frequency,
is listed for several polymers in Table 3.1. (When the FCP plots at different
frequencies were not parallel to one another, the FSF was computed from the
linear and parallel portions of the curves.)

Several attempts have been made to rationalize these data with varied
success. It was initially concluded that those polymers (e.g., PMMA, PS,
PVC) that exhibited a strong FCP frequency sensitivity also exhibited a
strong tendency to undergo heterogeneous deformation by crazing; negligible frequency sensitivity was associated with materials (e.g., nylon 66,
PVDF) that were craze resistant [34]. The generality of this postulate is
somewhat clouded by the fact that PC and PSF both experience craze
deformation [45] and yet experience FCP rates that are not sensitive to cyclic
frequency (Table 3.1).

A frequency sensitive change in any mechanical property may often be
traced logically to a time-dependent chemical effect involving an interaction
between the material and the test environment. For example, corrosion-
enhanced FCP behavior in metal alloys has been shown to depend strongly on
test frequency (i.e., test time) [1]. To evaluate this potential effect, FCP tests
were conducted on PMMA, PVC, and PC in dry nitrogen, laboratory air,
and water at several test frequencies. The results are given in Table 3.2 and
reveal that for the material–environment–frequency combinations ex-

TABLE 3.1   Frequency Sensitivity Factor (FSF) in
Polymers at Room Temperature[a]

Material	FSF	Material	FSF
PMMA	2.5–3.3	PC	1
PS	2.3	PSF	1
PPO/HIPS	2	Nylon 66	1
PVC	1.8	PVDF	1
CLPS	<1.5		

[a] See [21, 34–36].

TABLE 3.2   Effect of Mild Environments on FCP Response
in Polymers[a]

Material	Environment	Test frequency (Hz)	Effect
PVC	N_2, air	1, 100	None
PMMA	H_2O, air, N_2	10	None
PC	H_2O, air, N_2	10	Slight

[a] See [34].

amined, frequency sensitivity did not appear to be related to environmental effects [34]. Note that the slight environmental effect for the case of PC was found in a material that is essentially insensitive to frequency (see Table 3.1). Other material–environment combinations that have a more pronounced influence on FCP are discussed in Section 3.6.

Since the mechanical properties of polymeric solids are often sensitive to changes in strain rate, it is appropriate to determine whether a material's FCP frequency sensitivity may be related to changes in strain rate associated with different cyclic frequencies. In this regard, Knauss and Dietman [46] and Wnuk [47] introduced the concept of an effective stress intensity factor

$$\Delta K_{eff} = \Delta K_{elas}\, I(t)/I(o), \qquad\qquad (3.13)$$

where ΔK_{eff} is the effective stress intensity factor range, ΔK_{elas} the completely elastic stress intensity range, and $I(t)/I(o)$ the normalized creep compliance of the material. For an ideal elastic solid, $I(t)/I(o) = 1$ and $\Delta K_{eff} = \Delta K_{elas}$. For times greater than zero, the value of $I(t)$ is determined for a time equal to that necessary for a crack (moving at a certain rate) to traverse a characteristic distance of the material. According to Wnuk [48], this distance may be determined experimentally for each material. More recently, Wnuk [49] postulated that the FCP rate in a polymer could be given by the superposition of elastic and viscoelastic components with the form

$$da/dN = C_1(\Delta K/K_{Ic})^{n_1} + C_2(\Delta K/K_{Ic})^{n_2} f^{-1} \qquad (3.14)$$

The first term in Eq. (3.14) is rate independent while the second term possesses two rate-dependent quantities, test frequency f and the constant C_2, which represents the normalized creep compliance $I(t)/I(o)$. At low test frequencies, C_2 would become progressively greater than unity with the second term in Eq. (3.14) increasing in relative importance. Consequently, this relationship could predict the observed increase in FCP rate with decreasing test frequency found in PVC, PMMA, PS, and PPO/HIPS *if* time-dependent changes in the material compliance actually occurred. A similar conclusion was reached by Williams [50] in a more recent publication. It should be

TABLE 3.3   Frequency Induced Modulus Changes in Polymeric Solids[a]

Material	$E_{1\,Hz}/E_{0.1\,Hz}$	$E_{10\,Hz}/E_{1\,Hz}$	$E_{100\,Hz}/E_{10\,Hz}$	FSF
PVC ($M_w = 1.4 \times 10^5$)	0.99	1.01	1.01	2.3
PPO/HIPS	1.02	1.02	1.01	2.0
PS	1.02	1.03	1.02	2.2
PMMA ($M_w = 1.6 \times 10^6$)	1.10	1.09	—	2.6[+]
Nylon 66	1.01	1.00	0.99	1.0
PC	1.01	1.02	—	1.0
ABS	1.02	1.01	1.00	1.0

[a] See [41].

pointed out, however, that the frequency-sensitive materials (Table 3.1) are glassy in character at the test temperature with only a negligible variation in modulus E being reported over several decades of test frequency [51–53]. Even semicrystalline polymers show relatively small time-dependent changes in modulus below T_g [54]. To further clarify this point, additional tests were recently conducted for the purpose of determining the materials' elastic modulus during fatigue cycling between 0.1 and 100 Hz [41]. Measured values of E for several materials did not change to any significant degree (Table 3.3). Note for the case of PVC, PS, and PPO/HIPS that the measured values of E changed by only about 1% for each of several decade changes in cyclic test frequency. Clearly the recorded FSF values for these materials can not be rationalized on the basis of frequency-induced changes in elastic modulus. (Some correlation, however, is to be noted for the case of PMMA.)

It should be recognized that a change in test frequency alters not only the strain rate but also the number of loading cycles per unit time and the integrated time under load for each load excursion. As such, a study of strain rate effects on FCP behavior might better be examined by conducting fatigue tests at a fixed frequency but with different waveforms. For example, loading rates could be increased by switching from a positive sawtooth waveform to triangular, sinusoidal, negative sawtooth, and square waveforms in that order. Note, however, that the integrated area under these curves is not always the same, thereby making it impossible in every instance to isolate the effects of strain rate from the effects of possible creep on FCP behavior. Some waveform data for several different polymers are summarized in Table 3.4 with additional data shown in Fig. 3.7 for PC [55, 56]. It is clear that the FCP rates of these materials exhibit markedly different sensitivity to the character of the load wave profile. For example, Harris and Ward [57] found FCP rates in a vinyl urethane to be approximately six times slower with a square wave than with a triangular load wave profile. They argued that the improved fatigue response of this material with the square waveform was due to stif-

TABLE 3.4   Crack Growth Rates as a Function of Waveform at 1 Hz

Material	$\Delta K (MPa \cdot m^{1/2})$	$(da/dN) \times 10^4$ (mm/cycle)				
		∿	⊓	⊿	⋀	⟋
PVC[55]	0.72	1.17	1.41	1.02	1.08	0.8
PS[55]	0.77	—	21.6	17.8	14.5	13.9
PMMA[34]	0.83	—	19.2	7.8	8.2	7.8
Epoxy(A/E = 1.6)[55]	0.70	2.26	2.20	—	2.3	—
PC[56]	1–3	No Change		—	—	—
Vinyl Urethane[57]	1.25	—	17	—	100	—

fening and strengthening in the rubbery material associated with the very high loading rate of the square wave. A strong but opposite waveform effect was found for the case of the glassy polymers PMMA, PS, and PVC. For example, the FCP rate in PMMA associated with a square wave profile was 2.4 times greater than that associated with the triangular waveform [34]. Finally, no significant difference in FCP behavior was noted in PC [58] and the epoxy resin [55] with any type of waveform. Also, Arad et al. [59] found no difference in crack growth rates in PC when sinusoidal and triangular waveforms were used.

To summarize, the effect of load waveform on fatigue behavior has been examined for several polymers with the square wave load profile providing a strongly beneficial effect, no effect, or a decidedly deleterious effect on polymer fatigue response. Since the materials shown in Table 3.4 differ with regard to their relative viscoelastic response, different strain rate effects may be expected. However, as pointed out earlier, while the waveform experiments were designed to evaluate material response under different loading rates, the waveforms themselves possess different load-time integrated areas, thereby superimposing a creep effect on the fatigue process. Consequently, it is necessary to isolate load rate and load-time area changes from one another. Some preliminary data shown in Table 3.4 confirm the existence of a creep effect. For the same loading rate, square and negative sawtooth waveforms, crack growth rates were consistently higher in association with the square wave that possesses twice the load-time integrated area as compared with the sawtooth waveform. For PMMA, the fatigue crack growth rates were the same for all three triangular waveforms (same load-time integrated area) even though the loading rates were different. The importance of creep effects regarding the FCP response of PMMA is further demonstrated with a slightly different waveform experiment [60]. In this instance a nonsymmetrical square wave load profile was used with one full loading cycle taking place in 10 sec. In one sequence, a specimen experienced the maximum load for

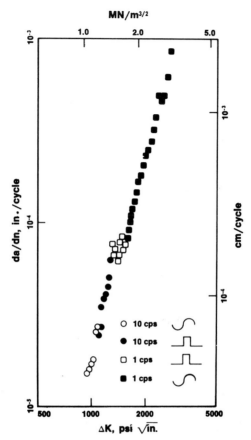

Fig. 3.7 Effect of cyclic wave form on FCP in polystyrene [Manson, Hertzberg, Kim, and Wu (56)].

0.2 sec and was completely unloaded for 9.8 sec, while in the other experiment the specimen was loaded for 9.8 sec and unloaded for only 0.2 sec. The crack growth rate associated with the latter load profile was 20 times greater than in the former case. Clearly, the FCP behavior in PMMA is dependent on creep effects to some significant degree. Beardmore and Rabinowitz [61] also reported a deleterious effect of enhanced creep during fatigue testing of unnotched PS samples in association with square wave testing. Also, Oberbach [62] reported for a polyamide resin that thermal failures in unnotched samples occurred at lower cyclic stresses and after fewer loading cycles when a square wave load profile was used as opposed to a triangular waveform. As a final note, the differences in crack growth rates in PS and PVC for the three different triangular waveforms (having equal load-time integrated area) reflect the fact that *both* loading rate *and* load-time integrated area must be considered when evaluating the FCP response of these two materials. Clearly, the FCP process in these materials is complex.

To this point, several explanations for fatigue crack propagation frequency sensitivity have been proposed and found wanting to various degrees. There remains for discussion one additional correlation that appears to hold the greatest promise for setting in order the complex body of fatigue data involving cyclic frequency as a test variable. A critical observation has been made that the frequency sensitivity factor (FSF) for a given polymer (Table 3.1) depends strongly on the frequency of movement of main chain segments (i.e., the jump frequency) responsible for generating the principal secondary transition peak (β peak) at a common test temperature (Fig. 3.8) [34, 41, 55, 63–65]. Values of jump frequency were estimated by extrapolating or interpolating existing data for the materials shown in Table 3.1 as follows: for PMMA, PC, PS, PVC, PVDF, and nylon 66 by McCrum *et al.* [52]; for PPO/HIPS by de Petris *et al.* [66], Stoelting *et al.* [67], and Eisenberg and Cayrol [68]; and for PSF by Butta *et al.* [69]. From the data presented, the greatest FCP frequency sensitivity occurred when the test frequency was close to the frequency of the β-process. Conversely, the β-frequency was far above the test range for the insensitive materials.

This correlation suggests a condition of resonance of the *externally* imposed test *machine* frequency with the material's *internal* segmental

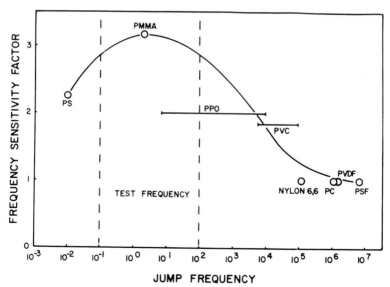

Fig. 3.8 Relationship between FCP frequency sensitivity and the room temperature jump frequency for several polymers [Hertzberg, Manson, Skibo (34)].

mobility corresponding to the β-peak. As a consequence, one finds a fascinating relationship between large- and small-scale deformation behavior, i.e., fatigue fracture and viscoelastic damping associated with the β-transition, respectively. A review of the polymer literature finds much precedent for such a correlation. During the past decade, it has become clear that a number of mechanical phenomena in polymers are closely related to the location (in terms of temperature and frequency) and other characteristics of the β-transition. As discussed by Oberst [70], Heijboer [71, 72], Boyer [73], and Vincent [74], impact strength often undergoes a transition in the region of the β-relaxation. Retting [75] and Bauwens [76] have shown that the yield stress of PVC is related to the relative position and intensity of the test conditions with respect to the β-transition. Broutman and Kobayashi [77] observed that the fracture energy of PMMA was directly related to both the α (glass)- and β (secondary)-transitions. Johnson and Radon [6, 7], have also shown that the time-to-failure in static tests with PMMA and PVC (at the ductile–brittle transition temperature) is approximately equal to the relaxation time of the β-process at the same temperature.

On the basis of the correlation shown in Fig. 3.8, one would expect the room temperature frequency sensitivity factor of PC, PSF, nylon 66, and PVDF to increase were it possible to excite these materials at test frequencies in the range of 10^6 Hz. Unfortunately, this could not be studied directly because of test machine limitations. However, since the segmental motion jump frequency varies with temperature, it is possible to choose a particular test temperature for each material that will bring the jump frequency into the cyclic frequency range permitted by the test machine. In this case, the frequency sensitivity would be maximized. Correspondingly, the frequency sensitivity of PMMA should be attenuated at test temperatures below ambient. Indeed, polysulfone and polycarbonate, two polymers that showed negligible frequency sensitivity at room temperature, each exhibited a maximum in the frequency sensitivity factor (>2.4) at temperatures of approximately 175 and $200°K$, respectively, corresponding to a jump frequency between 1 and 100 Hz for both materials (Figs. 3.9a and 3.9b) [64]. On the other hand, PMMA, which demonstrated a maximum frequency sensitivity factor at room temperature (jump frequency $\approx$ test frequency), responded to a lowering of test temperature with a considerable decrease in frequency sensitivity factor as the jump frequency became much lower than test frequency; at $150°K$ all frequency sensitivity was eliminated (Fig. 3.9c). Radon and Culver have also reported a lower FSF value in PMMA at temperatures below ambient [78].

It would appear then that the maximum sensitivity of FCP to frequency in engineering plastics occurs at a particular temperature unique to that material. In each case that temperature was related to a β-transition segmental jump frequency comparable to the test machine loading frequency. Of great significance, the overall frequency sensitivity for all the engineering plastics tested thus far has been shown to be singularly dependent on $T - T_\beta$, the difference between the test temperature and the temperature corresponding to the β damping peak within the appropriate test frequency range (Fig. 3.9d) [65]. As seen in Fig. 3.9d, the FSF parameter is maximized when the test temperature is equal to the temperature of the β-peak. [It is recognized

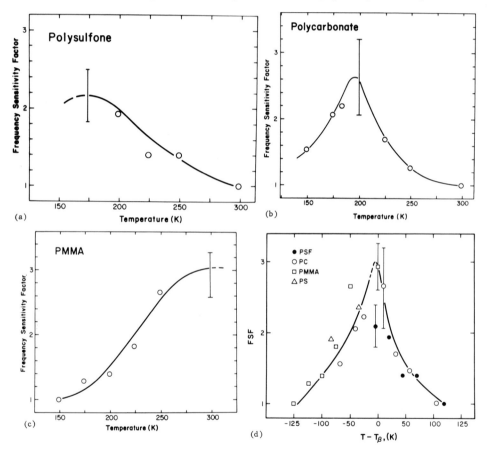

Fig. 3.9   Effect of temperature on frequency sensitivity in (a) polysulfone, (b) polycarbonate, and (c) poly(methyl methacrylate). [Reprinted from "The Effect of Temperature on the Frequency Sensitivity of Fatigue Crack Propagation in Polymers," by M. D. Skibo, R. W. Hertzberg, and J. A. Manson in *Advances in Research on the Strength and Fracture of Materials*, edited by D. M. R. Taplin, published by Pergamon Press Inc., 1978.]; (d) frequency sensitivity factor relative to normalized β-transition temperature, $T - T_\beta$. [Reproduced from R. W. Hertzberg, J. A. Manson, and M. D. Skibo, *Polymer*, 1978, **19**, 359, by permission of the publishers, IPC Business Press Ltd. ©.]

that the precise temperature associated with the β-process is itself a function of test frequency. We find, however, that the associated error in Fig. 3.9d (i.e., in the quantity $T - T_\beta$) is not significant]. Finally, Williams [50] has argued in somewhat related fashion that the extent of fatigue crack growth rate frequency sensitivity is dependent on the *magnitude* of tan δ at the β-peak as opposed to the quantity $T - T_\beta$ (Fig. 3.9d). In both instances, a strong connection has been established between gross mechanical properties (FCP response) and fine-scale viscoelastic response of a polymeric solid.

A physical interpretation of this correlation may be seen with the following model. Since the β-peak represents a region of maximum loss compliance, associated with considerable damping and energy dissipation, hysteretic heating should occur along with a localized temperature rise. In precracked samples utilized in FCP experiments, the maximum heat rise is restricted to the plastic zone near the crack tip; the bulk of the specimen experiences lower cyclical stresses and remains essentially at ambient temperature. Although heat transfer from the plastic zone to its cooler surrounding environment might limit the rate of crack-tip heating, fatigue testing at high frequencies should nevertheless produce a finite temperature rise at the crack tip. To wit, Altermo and Östberg [79] recorded a maximum increase in crack-tip temperature of up to 20° K in fatigue tests of PVC, PMMA, and PC at 11 Hz. With a significant increase in temperature, yielding processes in the material surrounding the crack tip should be enhanced and lead to an increase in the crack-tip radius. This greater radius of curvature at the crack tip should result in a lower effective ΔK. As this effective ΔK decreases, the fatigue crack growth rate is expected to decrease accordingly. In support of this model, high-frequency tests were performed on internally plasticized PMMA (large loss compliance) and PVC containing a 20% external plasticizing agent and resulted in a considerable amount of crack-tip heating [60]. Furthermore, the crack tip became visibly rounded whereupon stable fatigue crack growth ceased.

It has been shown then, that hysteretic heating on a *large scale* contributes to premature thermal failures during fatigue testing of *unnotched* samples while *localized* heating at the crack tip of a *precracked* FCP specimen leads to attenuated fatigue crack propagation rates. As pointed out in Section 2.2.2, the distinction between localized and generalized heating depends on the size of the hysteretically heated damage zone volume relative to the entire specimen volume. It might be possible, therefore, to experience *higher* FCP rates in a precracked sample at high test frequencies if the polymer possessed a high degree of damping. This special condition was found to exist in an impact-modified nylon 66 (HI-N66) [80, 80a]. This material is presumed to possess a very high degree of internal damping, based on the fact that it exhibited impact strengths more than an order-of-magnitude greater than those typical of unmodified nylon 66 [80b]. When HI-N66 was added to nylon 66, FCP rates decreased in proportion to the HI-N66 concentration but the relative ranking of the blends depended on the value of ΔK concerned. Interestingly, the FCP curves for all the blends (Fig. 3.10a) showed a change in slope at intermediate ΔK levels, the inflection point occurring at a *lower* ΔK level, the higher the HI-N66 content. An attempt was made to account for the shift in slope of the $da/dN - \Delta K$ plots as well as the relative ranking of each material in terms of yield strength and elastic modulus differences among the various nylon blends. To reflect yield strength differences, all crack length and associated ΔK values were adjusted by including a plastic zone size correction [80a]. These ΔK data were then normalized by the respective elastic moduli for each nylon 66 blend (see Section 3.8.7). The plots of $da/dN - \Delta K_{\mathrm{eff}}/E$ are shown in Fig. 3.10b and reveal HI-N66 (Zytel ST801) to be the most fatigue resistant material examined in that investigation. From the above computations, we conclude that the relatively

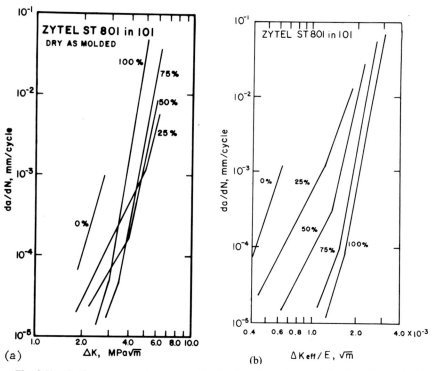

Fig. 3.10. Fatigue response in various blends of an impact modified nylon 66 (Zytel ST 801) in neat nylon 66 (Zytel 101). FCP rates plotted versus (a) $\Delta K_{\mathrm{applied}}$ and (b) $\Delta K_{\mathrm{effective}}/E$. [Skibo, Hertzberg, and Manson (80, 80a)].

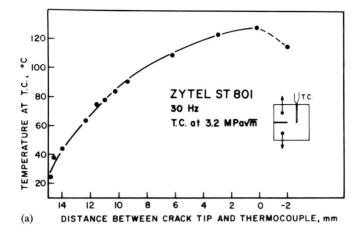

(a)

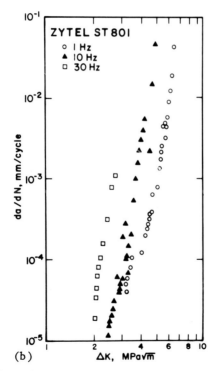

(b)

Fig. 3.11 Hysteretic heating induced temperature rise in moisture equilibrated impact modified nylon 66. (a) Temperature rise at thermocouple site along crack plane; (b) impact-modified nylon 66 FCP rates as a function of test frequency [Skibo, Hertzberg, and Manson (80)].

poor FCP response of the HI-N66-rich blends at high ΔK levels was due to lower yield strengths—giving rise to larger plastic zones than in the lean HI-N66 blends—and lower elastic moduli, the latter resulting in larger cyclic strains per unit stress intensity level.

This situation was aggravated by hysteretic heating—induced further reductions in these two properties. For example, a pure HI-N66 (moisture equilibrated at room temperature) was fatigue tested at 30 Hz and experienced a crack tip temperature of 125°C as measured by a thermocouple embedded along the crack plane (Fig. 3.11a). Note that considerable heating was taking place at the thermocouple site even when the crack tip was several millimeters away. By comparison, a crack tip temperature of 55°C [at $\Delta K = 3.7$ MPa·m$^{1/2}$] was measured when a dry as-molded sample of HI-N66 was fatigue tested at 10 Hz.

These observations of gross heating effects in precracked specimens strongly suggested that FCP rates in HI-N66-rich blends would *increase* with frequency. Recall that nylon 66 showed no FCP frequency sensitivity (Fig. 3.6b). As shown in Fig. 3.11b, FCP rates in pure HI-N66 increased with increasing test frequency from 1 to 30 Hz. Thus, this toughened nylon is the first polymer observed to exhibit higher FCP rates with increasing test frequency. It is seen then that the antipodal behavior of this material with that of PMMA, PS, or PVC reflects a different balance between gross hysteretic heating (which lowers the elastic modulus overall) and localized crack-tip heating (which involves crack-tip blunting). Presumably, materials like nylon 66, PC, and PSF at room temperature do not undergo even local crack-tip heating with the result that no FCP rate frequency sensitivity is observed.

3.5 EFFECT OF TEST TEMPERATURE ON FCP

Compared with FCP studies involving test frequency variations, very little information has been gathered to identify the effect of test temperatures on FCP behavior. As such, it is not yet possible to make fatigue crack growth rate predictions at nonambient temperatures using such methods as the time–temperature equivalency concept [81]. Furthermore, the limited amount of available data are not self-consistent since test methods differed among the reporting laboratories and different materials were examined that possessed different viscoelastic and deformation characteristics. For example, Kurobe and Wakashima [82, 83] found that macroscopic FCP rates in PMMA increased with decreasing temperature over a testing range of 50 to −10°C. Conversely, Radon and Culver [78] and Skibo [55] found that at 1, 5,

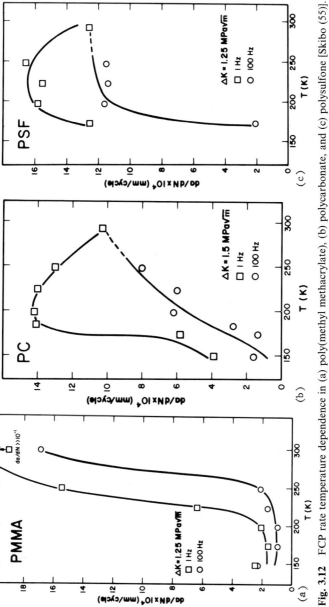

Fig. 3.12 FCP rate temperature dependence in (a) poly(methyl methacrylate), (b) polycarbonate, and (c) polysulfone [Skibo (55)].

and 100 Hz, FCP rates decreased monotonically with decreasing temperature. A sampling of the latter data is shown in Fig. 3.12 where the crack
growth rate at a given ΔK level is seen to decrease continuously from 22°C
(295°K) to −125°C (148°K) for both 1 and 100 Hz test frequencies. The conflicting trend in the data noted among these three investigations resulted

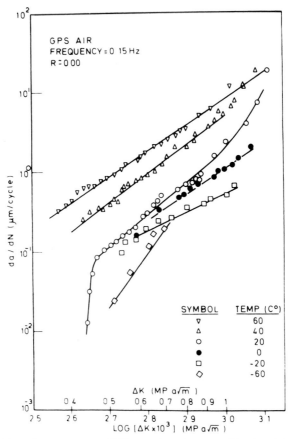

Fig. 3.13 Effect of temperature on FCP rates in polystyrene. [Reprinted from Y. W. Mai
and J. G. Williams, *J. Mater. Sci.* **14**(8), 1979, 1933, Chapman and Hall Ltd.]

Symbol	Temp (C°)
▽	60
△	40
○	20
●	0
□	− 20
◇	− 60

from the use of different experimental test procedures. The Radon, Culver, and Skibo data were generated under fixed load test conditions while the Kurobe and Wakashima experiments were conducted with constant displacement test equipment. Since the modulus of elasticity would be expected to rise with decreasing test temperature, the magnitude of the load associated with constant displacement test conditions (employed by Kurobe and Wakashima) would be expected to increase. Hence, specimens tested under fixed displacement control should experience a larger applied load range and associated higher FCP rates.

Fatigue crack growth rates in polystyrene under fixed loading test conditions also have been shown to decrease with decreasing test temperature. Mai and Williams [84] reported a 20-fold decrease in FCP rate at 0.15 Hz when the test temperature was reduced from 60 to −60°C (Fig. 3.13).

The FCP test temperature dependence of polycarbonate and polysulfone is much more complex than the behavior just described. Gerberich and co-workers [85, 86] showed that fatigue crack growth rates at 1 Hz *increased* with decreasing temperature down to −50°C and then *decreased* with a further drop in test temperature. By plotting the ΔK level necessary to produce a crack growth rate of some specific value as a function of temperature, they showed that ΔK was a minimum at about −50°C for both of these materials (Fig. 3.14). K_{Ic} was also shown to be a minimum at this temperature. Skibo [55] examined the fatigue behavior of PC and PSF and found a similar fatigue response to test temperature at 1 Hz but not at 100 Hz. Some of these data are given in Figs. 3.12b and 3.12c where the FCP rate is shown as a function of temperature for a fixed ΔK value. The maximum crack growth rate at 1 Hz is found at about −70°C in PC and about −50°C in PSF. On the other hand, the FCP rates in these materials decreased continuously with lower temperatures at a test frequency of 100 Hz, much like the results reported for PMMA (Fig. 3.12a). To confuse the situation still further, Radon and Culver [78] reported that the FCP rate in PMMA increased as the temperature decreased from 40 to 10°C, whereupon growth rates decreased with a further lowering of test temperature. To summarize these experimental findings, PMMA and PS fatigue specimens exhibited monotonically decreasing FCP rates with decreasing test temperature at test frequencies from 5 to 100 Hz (except for an intermediate crack growth rate maximum at 20 Hz as reported by Radon and Culver); Gerberich and co-workers and Skibo revealed maxima in FCP rate at intermediate temperatures in PC and PSF at 1 Hz while Skibo found monotonically decreasing growth rates with decreasing temperatures at 100 Hz in PC.

Martin and Gerberich [85] have attempted to rationalize the intermediate temperature FCP rate maximum found in PC in terms of the viscoelastic response of this material. They assumed that adiabatic heating occurred near the fatigue crack tip and computed that a local temperature rise of 100°C

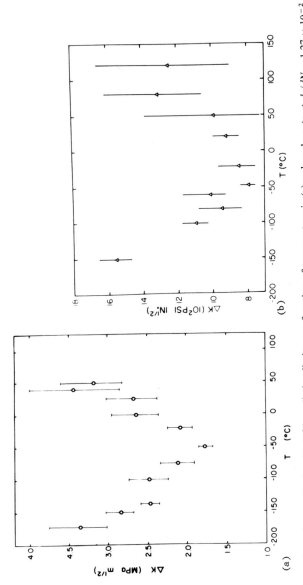

Fig. 3.14 Stress intensity amplitude with 95% prediction limits as a function of temperature in (a) polycarbonate at $da/dN = 1.27 \times 10^{-2}$ mm/cycle [G. C. Martin and W. W. Gerberich, *J. Mater. Sci.* **11**, 1976, 231, Chapman and Hall Ltd.]; (b) polysulfone at $da/dN = 7.5 \times 10^{-3}$ mm/cycle [Wann, Martin, and Gerberich (86)].

took place [87]. After averaging the integrated damping loss over this 100°C temperature rise, they found that the minimum in K_{Ic} values and maximum in FCP rates at $-50°C$ could be correlated with the magnitude of the secondary loss. Though the approach taken by Martin and Gerberich holds promise, there are some difficulties regarding its application to PC data as done in this instance. Principally, the present authors question some of the assumptions made in the computation that led to the conclusion that a 100°C crack-tip temperature rise can occur in PC when tested at 1 Hz. In fact, Attermo and Östberg [79] reported an increase in crack-tip temperature in PC of less than 30°C at 11 Hz. Consequently, a much smaller crack-tip temperature increase would be expected at a test frequency of only 1 Hz. Part of the difficulty of equating the computed crack-tip temperature elevation with the actual crack-tip temperature measurement is the fact that the value of crack velocity used to compute the temperature increase was several orders of magnitude higher than the actual fatigue crack growth rate [87]. In fact, the use of a much smaller crack velocity would have resulted in the computation of a much smaller crack-tip temperature rise. In considering this point, Gerberich [88] has suggested the possibility that the fatigue crack velocity might be larger than one might expect, based on average macroscopic growth rate data. For example, one might speculate that the crack grows discontinuously and in rapid short bursts with associated high velocity. Indeed, there are situations where this does occur in PC as well as in other several other polymeric solids [45] (see Chapter 4). However, no such discontinuous cracking has been found in the ΔK regime investigated by Martin and Gerberich [85] and Skibo *et al.* [45, 55].

On the basis of these incomplete exploratory results, clarification of the temperature dependence of FCP must await further experimentation.

3.6 EFFECT OF ENVIRONMENT ON FCP

Limited investigations have been carried out to identify the effect of well-characterized test environments on FCP in polymeric solids. To further complicate the situation, it should be recognized that most prior fatigue tests were performed under unspecified laboratory conditions. As devilishly stated by Marshall and Williams [89], "In these polluted times who knows what constitutes laboratory conditions?" Some earlier studies employed the surface work parameter $\mathscr{T}$ [see Eq. (3.12)] to correlate FCP rates in natural rubber as a function of ozone concentration in air [24, 90–93]. It was found that cracks grew at lower values of $\mathscr{T}$ in the presence of ozone as opposed to air. More recently, the stress intensity factor has been used to correlate

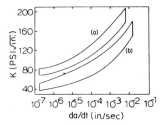

Fig. 3.15   Crack extension rate *da/dt* in polyethylene as function of stress intensity factor. Band A: melt flow index 7, in methanol. Band B: melt flow index 20, in methanol [Marshall, Culver, and Williams (95)].

environmental crack growth and solvent craze formation in semicrystalline and amorphous polymers under both static and cyclic loading conditions [94–99]. For example, Marshall *et al.* [95] found crack growth rates in polyethylene in methanol to be controlled by the prevailing stress intensity factor value at the advancing crack tip (Fig. 3.15). Of particular interest, the rate of crack extension at a given stress intensity factor level decreased with decreasing melt flow index (increasing molecular weight).

These results were subsequently used as a reference point for an examination of environmental effects of methanol on FCP in polyethylene. El-Hakeem and Culver [100] noted that fatigue crack growth rates were greatly enhanced when the crack tip was immersed in a methanol bath as opposed to laboratory air. In fact, environmentally assisted FCP terminated at a ΔK level that was below that required to *initiate* fatigue crack extension in air, i.e., below the FCP threshold level ΔK_{th}. In a related study, Marshall *et al.* [94] examined environmentally assisted craze initiation and growth in PMMA. A threshold stress intensity level was defined below which craze growth did not occur. In addition, two different craze growth mechanisms were identified at low- and high-stress intensity factor levels, respectively, and related to variations in the nature of liquid flow to the craze tip.

Martin and Johnson [101] examined the effect of molecular weight and vapor environment on the fatigue life of PVC films. Tests were conducted in dry nitrogen and nitrogen saturated with both water and ethanol vapors. Fatigue testing in water vapor at 40°C led to a general plasticization of the PVC films with the complex modulus decreasing by a factor of three and the value of tan δ increasing from 0.046 to 0.11 after only 20 min of testing at 3.5 Hz. By contrast, both complex modulus and tan δ values remained essentially constant during cyclic loading in ethanol vapors until craze nucleation or fracture occurred. The effect of both molecular weight and ethanol vapor on fatigue life in PVC films is seen in Fig. 3.16. These results show that there is an approximate eight- to tenfold decrease in fatigue life when ethanol vapors are added to a nitrogen environment. Also, the fatigue life of the PVC

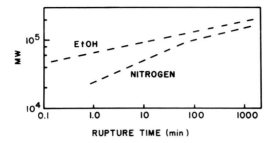

Fig. 3.16 Effect of molecular weight and ethanol vapor on fatigue life in PVC films. [J. R. Martin and J. F. Johnson, *J. Appl. Polym. Sci.*, **18**, 1974, 3227. Reprinted by permission of John Wiley & Sons, Inc. © 1974.]

films increased by four orders of magnitude when M_w was increased sixfold. One may conclude from these data that the deleterious effect of an aggressive environment on fatigue life can at least be partially offset by using a higher M_w polymer. (This is consistent with observations of stress cracking in polymers under sustained loading conditions.) That is to say, a higher M_w material should be specified for a given engineering application when aggressive environment service conditions are anticipated. This behavior closely parallels the findings of Marshall *et al.* [95] where the environmentally assisted stress cracking rate in polyethylene decreased with decreasing melt flow index (increasing molecular weight). Martin and Johnson [101] speculated that the ethanol vapors enhanced chain mobility, thereby hastening craze formation and reducing the time for chain entanglements to reach their maximum extension. Without the presence of ethanol, they argued that the rate of chain motion would decrease and the fatigue life increase as a consequence.

Mai [102] examined the fatigue crack propagation behavior of PMMA in ethanol and carbon tetrachloride at different cyclic frequencies. The data shown in Fig. 3.17 reveal the influence of a CCl_4 environment and different cyclic frequencies on FCP response in this commercially important material. As expected, crack growth rates increased with decreasing test frequency with a frequency sensitivity factor of 20 being found at lower ΔK values. (In this study, crack growth rates were compared with values of ΔR, a measure of crack extension resistance that is proportional to ΔK^2.) Note that this frequency sensitivity is much higher than that obtained from tests conducted in laboratory air (see Table 3.1). This much higher FSF value at low ΔR levels suggests that the kinetics of fatigue crack growth in this ΔK regime are strongly influenced by the test environment. Note that at higher ΔR values, the slope of the curve decreases along with the FSF value that is still higher (about ten) than the laboratory air value (about three). The decreasing influence of test environment on FCP at high ΔK levels is similar to the re-

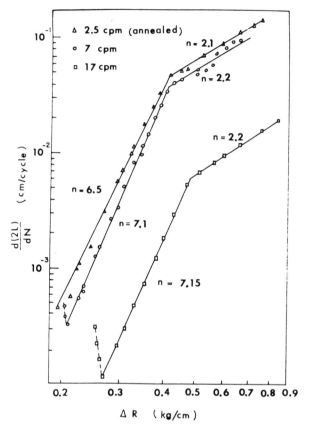

Fig. 3.17   Fatigue crack propagation behavior of PMMA in 99% CCl_4. [Reprinted from Y. W. Mai, *J. Mater. Sci.* **9**, 1974, 1896, Chapman and Hall, Ltd.]

sponse observed in metal alloys and suggests that the greater degree of mechanical damage associated with fatigue cracking at larger ΔK levels tends to overshadow the aggravating effect of test environment.

Recently completed fatigue crack propagation studies of polystyrene conducted in corn oil and detergent (Adinol) environments at 0.15 Hz have also revealed a two-stage cracking response (Fig. 3.18) [84]. The slope of the da/dN–ΔK plot decreased from 12 to 14 at low ΔK levels to about 2 at high ΔK values, again suggesting a greater role of environment at low ΔK levels in controlling fatigue crack growth behavior. Overall, these two liquids caused the FCP rate in polystyrene to increase by more than 100-fold over a temperature range from -20 to 60°C. Mai and Williams [84] further reported that the stress for the onset of crazing and the fracture toughness decreased in the presence of these two liquids. By contrast, Mai [102] reported that the

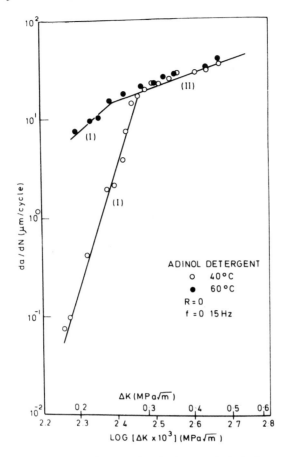

Fig. 3.18 Fatigue crack propagation behavior of polystyrene in Adinol detergent. [Reprinted from Y. W. Mai and J. G. Williams, *J. Mater. Sci.* **14**(8), 1979, 1933 Chapman and Hall, Ltd.]

fracture toughness in PMMA increased by about 60% in the presence of ethanol and carbon tetrachloride. It would appear then that the effect of environment on K_c and the stress for crazing depends on the specifics of the material–environment system in question.

As part of some preliminary studies, Manson *et al.* [21] reported an effect of water on FCP response in polycarbonate (Fig. 3.19). Note that the material undergoes a transition in its relative fatigue performance; below a ΔK level of about 2.2 MPa m$^{1/2}$, FCP rates are lower in dry nitrogen while above 3.3 MPa m$^{1/2}$, the material exhibits slower crack growth rates in water. Further studies are needed to clarify the meaning of this transition in FCP behavior. For example, tests conducted at different cyclic frequencies might lead to a shift in the position of the transition.

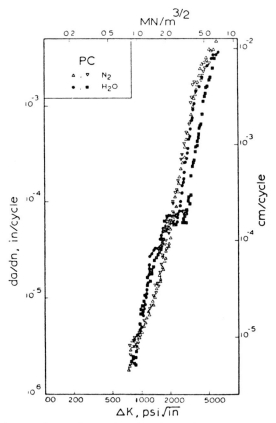

Fig. 3.19   Fatigue crack growth rates in polycarbonate in water (●, ■) and dry nitrogen
(△, ▽). [Reprinted with permission from *Crit. Rev. Macromol. Sci.*, 1(4), 1973, 433. Copyright
The Chemical Rubber Co., CRC Press, Inc.]

3.7   EFFECT OF MEAN STRESS AND LOAD
HISTORY ON FCP

A number of studies have been conducted to evaluate the effect of mean
stress on the rate of fatigue crack propagation at different ΔK levels. It is
confusing to find that a number of different relationships between mean stress
and FCP rate have been proposed for both polymers [21, 32, 33, 43, 59,
103] and metals [104, 105]. Unfortunately, many of these formulations have
been found wanting to one degree or another. For example, Arad *et al.* [32]

proposed that fatigue crack propagation rates could be related to the crack-tip stress intensity factor with a relationship of the form

$$da/dN = \beta \lambda^n, \tag{3.15}$$

where β is a function of loading frequency, material properties, and environment, n a constant, and $\lambda = K_{max}^2 - K_{min}^2$. Since λ is also equal to the quantity $2\Delta K\, K_{mean}$, Eq. (3.15) bears a strong resemblance to the equation

$$da/dN = CK_{mean}^{2.13} \Delta K^{2.39}, \tag{3.16}$$

which is based on a regression analysis of fatigue data from PMMA samples [33]. In either case, the fatigue crack growth rate at a constant ΔK level is seen to increase with increasing mean stress (or increasing stress ration R). Such behavior parallels that found in numerous metallic alloy systems [1].

The general applicability of Eq. (3.15) can be examined *for the case of PMMA* by recomputing the data of Mukherjee and Burns [33] in terms of λ and comparing with the results of Arad *et al.* [103] (Fig. 3.20). It is seen that a log–linear relationship between da/dN and λ does exist at a given frequency for this material. Other investigators have been encouraged by these findings

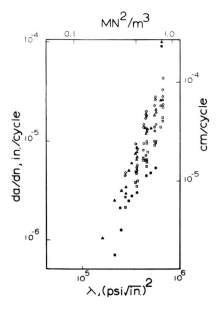

Fig. 3.20 Fatigue crack growth rate in PMMA as a function of λ ($= K_{max}^2 - K_{min}^2$). Data symbols refer to cited results from original source. [Reprinted with permission from *Crit. Rev. Macromol. Sci.* **1**(4), 1973, 433. Copyright The Chemical Rubber Co., CRC Press, Inc.]

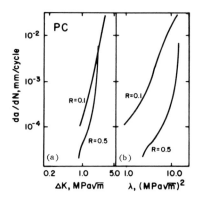

Fig. 3.21   Fatigue crack propagation in polycarbonate at two mean stress levels as a function of (a) ΔK and (b) λ. Note the superior correlation with ΔK. [Reprinted with permission from *Crit. Rev. Macromol. Sci.* **1**(4), 1973, 433. Copyright The Chemical Rubber Co., CRC Press, Inc.]

and have adopted the λ parameter in making FCP comparisons whether or not mean stress is varied. We believe that widespread adoption of the λ parameter is premature. Specifically, there are reports of fatigue crack propagation data at varying mean stress levels that can not be normalized with the λ term. For example, it was reported earlier [21] for most ΔK levels examined, that the crack growth rate in polycarbonate *decreased* with increasing mean stress level. In fact, it was found that the fatigue data were better correlated by using the ΔK parameter rather than λ (Fig. 3.21). Most recently, Mai and Williams [84] reached the same conclusion in their investigation of mean stress effects on FCP behavior in polystyrene. Specifically, when crack growth data were plotted versus ΔK, da/dN values increased by about two- to threefold at a given ΔK level when the load ratio R increased from 0 to 0.6. (Similarly, Sauer *et al.* [106] found that the unnotched fatigue life of PS samples decreased at a fixed stress range when the mean stress was increased.) On the other hand, da/dN values decreased by almost two orders of magnitude at a given λ value when the mean stress was changed by the same amount. The sharp *decrease* in da/dN with increasing R at a constant λ arises since ΔK decreases according to the relationship

$$\Delta K = \{[(1 - R)/(1 + R)]\lambda\}^{1/2}. \tag{3.17}$$

To explore this dichotomy in more detail, some exploratory experiments were conducted by Skibo [60]. These results are summarized in Table 3.5 along with observations by others. As before, some polymers exhibited an increase in crack growth rate at a fixed ΔK level with increasing R. As mentioned earlier, such behavior is consistent with the response of metallic alloy systems. On the other hand, the list of polymeric solids that exhibited lower crack growth rates with increasing R values was expanded.

TABLE 3.5   Effect of Mean Stress on FCP in Various Polymers

Material	Condition	Change in da/dN as R^a increases from 0.1 to 0.5	Reference
PMMA	Commercial sheet High M_w	Increase	[32, 33, 60]
PMMA	High M_w; cast	Increase	[60]
PVC	Molded; M_w—96,000	Increase	[60]
PVC	Molded; M_w—141,000	Increase	[60]
PVC	Molded; M_w—220,000	Increase	[60]
Epoxy	Amine/epoxy—1.8/1	Increase	[60]
Epoxy	Commercial grade	Increase	[107]
PVDF	Extruded	Increase	[60]
Nylon 66	Extruded	Increase	[60]
HI-N66	Extruded	Increase	[60]
Polystyrene	Extruded	Increase	[60, 84]
PVC	Commercial sheet extruded	Decrease	[60, 108]
PC	Commercial sheet extruded	Decrease	[21]
ABS	Commercial sheet extruded	Decrease	[60]
PSF	Commercial sheet extruded	Decrease	[60]
PPO/HIPS	Commercial sheet extruded	Decrease	[60]
LDPE	Commercial sheet	Decrease	[60, 109]

^{a} R = minimum load/maximum load.

The surprisingly beneficial influence of mean stress on fatigue crack propagation behavior is not without precedent, however. Some 40 years ago, researchers found that the fatigue life of rubber could be enhanced dramatically when the minimum strain in the fatigue cycle was increased, even if the loading *range* was kept constant. Cadwell *et al.* [110] reported that "... . rubber under linear vibrations exhibits a minimum dynamic fatigue life in the region where $L_{max} = L_0$, that is, where the return stroke brings the sample back to a condition of zero strain." Their data are shown in Fig. 3.22 and clearly reveal an increase in specimen fatigue life with increasing percentage minimum strain up to some maximum value. More recent fatigue studies, employing the fracture energy parameter T [see Eq. (3.12)], have shown that dramatic improvements in fatigue life corresponding to much lower fatigue crack growth rates at a given level of ΔT have been noted in strain crystallizable rubbers [24–26] (Fig. 3.23) and nylon fibers [111]. This marked improvement in fatigue resistance has been attributed to permanent strain-induced crystallization that suppressed crack extension [28, 112].

Apparently, mean stress effects on the fatigue response of polymeric solids are dependent on conflicting processes. On one hand, there may be a tendency for higher crack growth rates with higher mean stress intensity levels (as-

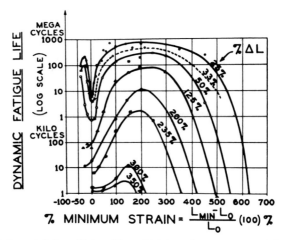

Fig. 3.22   Enhanced fatigue life in rubber with increasing percentage minimum strain up to some maximum value. [Reprinted with permission from *Ind. Eng. Chem. Anal. Ed.* **12**(1), 1940, 19. Copyright by the American Chemical Society.]

suming a constant stress intensity range) as a result of more creep crack extension associated with higher K_{max} levels and also because the maximum stress intensity factor approaches the critical limiting value for the material (i.e., K_c). Alternately, there may be a strain-induced modification of the polymer substructure, thereby making the fatigue process more difficult and resulting in lower fatigue crack growth rates. Regarding the latter premise, our studies have also shown that the apparent fracture toughness of many

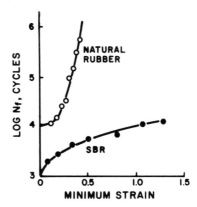

Fig. 3.23   Dependence of fatigue life on minimum strain in natural rubber ($\bigcirc$) (which strain crystallizes) and SBR ($\bullet$) (which does not crystallize). Maximum strain 2.5 for all points. [Reprinted from G. J. Lake and P. B. Lindley, Conference Series No. 1, Institute of Physics and Physical Society, London, 1966, p. 176 with permission from the Institute of Physics.]

polymers (defined at the maximum stable fatigue crack length) *increased* with increasing mean stress level. This totally unexpected finding may be due to permanent cyclically induced reconditioning of the polymer crack-tip substructure. Therefore, the strain energy that is available for crack extension may be consumed during elastic and plastic deformation and also diverted into providing for entropic changes within the damage zone. That is, some energy otherwise targeted for the fracture process may be consumed in changing the molecular conformation near the crack tip to a more ordered state. In summary, the FCP rate sensitivity of a polymeric solid to mean stress will depend on the relative importance of the competing factors discussed above.

Before closing this discussion, it is appropriate to examine these mean stress fatigue data in light of the generalized fracture theory proposed by Andrews [113]. Andrews proposed that the fracture energy parameter T could be given by

$$T = T_0\{C/[C - f(\beta)]\}, \tag{3.18}$$

where T is the total energy expended by the solid to cause unit area of crack growth, T_0 the energy expended in a perfectly elastic solid, C a function of strain, and β the hysteresis ratio. The hysteresis ratio reflects how much energy is lost within the hysteresis loop associated with inelastic deformation. It may be shown that if β is large, the value of T needed for fracture must increase. Correspondingly, it would be expected that FCP rates would decrease when β is increased. This is analogous to a reduction in crack growth rate with increasing fracture toughness values as predicted by Wnuk [49] [recall Eq. (3.14)]. In fact, a consistent trend has been shown between the polymer fracture toughness and the ΔK level necessary to drive a fatigue crack at a certain velocity (see Section 3.8.7).

Proceeding further, Andrews' fracture theory may be employed to rationalize the complex dependence of FCP on mean stress. For the case of elastomers, Andrews has shown that the value of β increases sharply with strain to a limiting value. Therefore, when a high mean stress is imposed on the specimen, the associated large strains will introduce a large value of β. From Eq. (3.18), a larger value of T would be needed to drive the crack. Alternatively, for a given ΔT (analogous to a given ΔK level), the fatigue crack growth rate would decrease with increasing load ratio R. Since the material retains a large strain down to small stress levels, then even a small positive stress will maintain a relatively large strain (Fig. 3.24). Therefore, the crack would remain blunt upon unloading to σ_{min} and not be sharpened in preparation for the next loading cycle. This would be expected to lead to reduced crack growth rates.

In general, we speculate that the resistance to FCP is related to β, which in

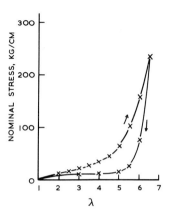

Fig. 3.24 Hysteresis loop in strain crystallizing natural rubber showing retention of large strain at small positive stress [Andrews (112)].

turn may be dependent on mean stress. For example, if $\beta \approx 0$, as in the case of an ideally elastic material, then an increase in mean stress would not be expected to alter significantly the crack-tip profile as was the case in strain crystallizable rubbers. Therefore, a higher mean stress should serve to increase K_{max} near K_c and promote faster crack growth rates. On the other hand, if the material experiences hysteretic losses, then a high mean stress should increase β so as to slow the crack advance rate by crack-tip blunting and/or strain-induced orientation hardening.

Since many engineering components experience complex loadings during their service life, it is appropriate to determine whether fatigue crack growth rate information generated under constant load amplitude testing conditions may be used to predict component cracking response under more complex load spectra conditions. Previous studies of this type have already been reported in various metallic alloys [1]. Of particular importance, it has been found that the introduction of a single overload or short cluster of overload cycles results in a temporary attenuation of the subsequent crack extension rate below that expected at a particular ΔK level. This delay in crack growth in certain cases can lead to prolonged service life. Pitoniak, Grandt, and co-workers [114–116] have conducted similar overload studies in PMMA and PC and found similar fracture phenomena. For example, they found when PMMA was cycled at a ΔK level of 0.5 MPa m$^{1/2}$ and then overloaded five times at 1.0 MPa m$^{1/2}$, that the crack arrested for 35,000 cycles before resuming its growth when $\Delta K = 0.5$ MPa m$^{1/2}$. In connection, with this crack arrest condition, optical interference fringe patterns were obscured on the fracture, signifying crack surface interference. Similar results were reported for tests conducted with PC, though the extent of crack growth delay in PC and PMMA was found to be less extensive than that reported for an aircraft grade aluminum alloy.

In a recent publication, Mai [116a] also reported fatigue crack growth retardation in PS. In this preliminary study it was found that the extent of cyclic delay increased with increasing overload ratios and that crack growth delay was due primarily to the overload induced formation of multiple crazes in the vicinity of the crack tip, which reduced the effective stress intensity factor. (See also Chapter 5 for similar effects discussed for composite materials.)

Though it may appear that deliberate overloading of polymeric solids would enhance component life under certain conditions, it should be pointed out that these same overload cycles may shorten fatigue life for different specimens and test conditions. Therefore, much more work is needed before the fatigue life can be predicted for components that would be expected to experience complex load histories.

3.8 EFFECTS OF MOLECULAR PROPERTIES AND MORPHOLOGY ON FCP

Attention is now given to an examination of those material variables that influence polymer fatigue crack propagation behavior. Such variables include molecular weight, molecular weight distribution, degree of cross-linking, amount of crystallinity, plasticizer content, static mechanical properties, composition, and thermal history. (The influence of second phase additions on polymer FCP response is discussed in Chapter 5.) Since these variables are known to be of great importance in determining many kinds of material behavior, from the viscosity in dilute solution of a polymer of theoretical interest to the impact strength in an engineering plastic, one would expect that fatigue properties should respond also to such variables in related fashion.

In a study designed to establish the range of crack propagation parameters in engineering plastics subjected to fatigue loading, Hertzberg et al. [117] noted dramatic differences in behavior among polymers possessing a wide range of structures and morphologies. The materials examined ranged from crystalline polymers such as nylon 66 to amorphous, glassy polymers such as PMMA and PC and a complex multicomponent polymer, acrylonitrile-butadiene-styrene copolymer (ABS). Results given in Fig. 3.3 revealed much larger differences in material behavior than that found in comparisons made among different metallic alloys. At a given ΔK level, for example, crack growth rates among these materials varied by three to four orders of magnitude.

Based on additional studies [21, 40] conducted on another group of polymers, ranging from amorphous PS to highly crystalline PVDF, several general trends in FCP behavior became clearly evident. First, regardless of the polymer structure and composition, a reasonably linear log ΔK–log(da/dN) relationship was found for all of the polymers examined thus far. Though differences are noted in the Paris growth rate exponent m [Eq. 3.10)] among the materials examined, such variations appear minor when compared with differences in the preexponential constant A. From previous discussions in Sections 3.4–3.7, the constant A is seen to vary with test variables such as frequency, temperature, environment, and mean stress level. What remains to be explored is the extent to which material variables impact on the magnitude of A. Certainly, a single structure–property relationship could not be reasonably expected. Even with multiple explanations on a case-by-case basis, the existence of anomalies should not be surprising, especially at very low and very high values of ΔK. Unfortunately, little work has been reported which describes the fatigue behavior of thoroughly characterized polymeric solids as compared with careful documentation of microstructural influences on the FCP response of metallic alloys. This apparent lack of interest in generating such information is attributable, in part, to the previously limited use of polymeric materials in load-bearing structural applications. In addition, a thorough characterization of polymer structure, composition, and viscoelastic properties is both difficult and tedious and often not within the laboratory capability of the designer or component fabricator.

3.8.1 Nature of Fatigue Damage

Before considering the influence of specific material variables on FCP behavior, it is useful to describe some general features associated with the fatigue and fracture process. From Section 3.2, the presence of a crack in a component under load creates a stress concentration in the vicinity of the crack tip. When the associated stress maxima in this region exceed the material's yield strength, a zone of damage is formed immediately ahead of the crack tip. One estimate of the size of this damaged zone [Eq. (3.5)] finds the radius of a cylindrically shaped zone to be proportional to the square of the stress intensity factor and inversely proportional to the square of the material's yield strength. Since the shape of the crack-tip plastic damage zone in polymeric solids very often assumes a more lens-shaped configuration, especially in connection with the formation of crazes, a number of investigators have adopted the Dugdale plastic strip model to better describe the crack tip damage zone [38, 45, 108, 118–120]. The Dugdale model supposes that the plastic strip is in the plane of the crack and bears a uniform stress equal to the yield strength of the material [121, 122]. Also, the model is concerned with conditions of plane stress where $\sigma_z = 0$ (see Fig. 3.2). As such, the model describes the condition associated with craze formation at a crack tip in that

    (1)   the craze assumes the shape of a narrow plastic strip;
    (2)   the craze is located in the plane of the crack;
    (3)   there is a uniform stress acting across the craze (i.e., the craze stress);
and
    (4)   the stress in the z direction of the craze is essentially zero.

The latter condition arises even though the overall stress conditions in the sample may be nominally plane strain. That is, the through-thickness stress is dissipated through hole formation within the craze. The Dugdale model describes the crack-tip plastic zone as a thin planarlike region with a total length

$$L \approx \tfrac{1}{8}\pi K^2/\sigma_{ys}^2,　(3.19)$$

where L is the distance from crack tip over which damage has occurred, K the stress intensity factor, and σ_{ys} the material yield strength. Note that the form of Eq. (3.19) is essentially the same as that given in Eq. (3.5). Proceeding further, the extent to which this damage zone stretches open is maximized at the crack tip and given by

$$\delta = K^2/\sigma_{ys} E,　(3.20)$$

where δ is the maximum damage zone opening displacement, E the modulus of elasticity, K the stress intensity factor, and σ_{ys} the material yield strength.

    With regard to the fatigue crack propagation behavior in polymeric solids, it is necessary to understand the process(es) by which the material contained within the damage zone breaks down. In the case of a craze zone, breakdown is envisioned to take place by cumulative rupture of the craze fibrils and coalescence of the microvoids within the porous craze zone. The gradual rupturing of the craze fibrils is believed, in turn, to be caused by a viscoelastic process involving cyclic stress-induced disentanglement of molecules linked together in each fibril [120]. A more detailed analysis of this damage accumulation process is given in Section 4.2.2 and discussed also in Sections 3.8.2 and 3.8.4.

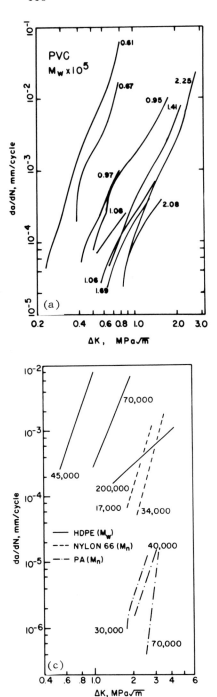

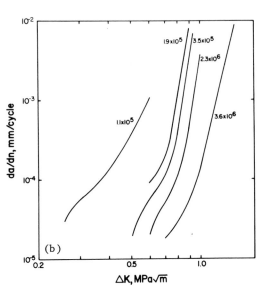

Fig. 3.25 Enhancement of fatigue crack propagation resistance with increasing molecular weight in (a) poly(vinyl chloride) $M_w \times 10^5$ [Rimnac, Manson, Hertzberg, Webler, and Skibo (128)]; (b) poly(methyl methacrylate) [Kim, Skibo, Manson, and Hertzberg (130)]; (c) high density polyethylene (—, M_w), nylon 66 (---, M_n) and polyacetal (--, M_n) [de Charentenay, Laghouati, and Dewas (131)].

3.8.2   Effect of Molecular Weight and Molecular Weight Distribution on FCP

As mentioned above, molecular weight (M) is a property characteristic of polymers, but not metals or ceramics, and hence often neglected. In addition, it is also not always known or easy to determine in commercial specimens. Studies by Sauer *et al.* [123–126] of fatigue lifetimes in unnotched specimens of polystyrene revealed a marked improvement in resistance to cyclic damage when M increased (see Section 2.6). A similar improvement in FCP resistance with increasing M has been reported in PVC [127, 128], PA [129], nylon 66, [129], PMMA [130], and PE [131]. A most striking observation is that there is a continuous shift to lower fatigue crack growth rates with increasing M for each of these materials (Fig. 3.25). This trend is emphasized for the case of PVC by comparing da/dN for different values of M at an arbitrarily chosen value of ΔK (0.7 MPa m$^{1/2}$). The results shown in Fig. 3.26 indicate a three orders-of-magnitude shift in da/dN values over the range of M tested with the major influence being found at lower values of M; indeed, the relationship is exponential with da/dN varying with $\exp(1/M)$ (Fig. 3.27).

The question of how M affects FCP behavior may be illuminated by considering a suggestion made some time ago in the light of new evidence about the structure and behavior of crazes, which, at least in PVC and PMMA, precede the growth of microscopic cracks. Berry [132] suggested that a craze could be stable only if the molecules were long enough to be firmly anchored at each internal face of a craze. As shown by Kausch [133],

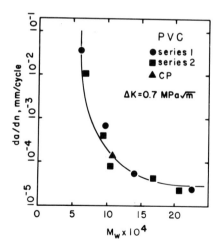

Fig. 3.26 Fatigue crack propagation rates in PVC at $\Delta K = 0.7$ MPa $\cdot$ m$^{1/2}$ as a function of M_w: ●, series 1; ■, series 2; ▲, CP [Rimnac, Manson, Hertzberg, Webler, and Skibo (128)].

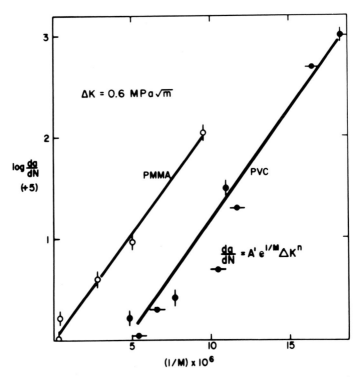

Fig. 3.27 Functional relationship between FCP rates in PMMA and PVC and reciprocal of molecular weight at 10 Hz for $\Delta K = 0.6$ MPa $\cdot$ m$^{1/2}$ [Kim, Skibo, Manson, and Hertzberg (130)].

such a simple model of chains extended singly or in bundles across a craze is physically unrealistic in that the high degree of orientation implied is not feasible. However, from the evidence developed by Kambour [134], Kusy and Turner [135, 136], Weidmann and Döll [137], and Wellinghoff and Baer [138], it is clear that only molecules longer than some characteristic length related to the spanning of a craze can form an entanglement network that can effectively resist crack propagation. That is, the craze is spanned by an effective entanglement network involving several molecules, rather than by individual extended chain molecules. We propose that cyclic loading tends to disentangle whatever network exists, and that this disentanglement is easier at lower M. Furthermore, there may be positive contributions from enhanced orientation hardening with the higher M species.

The profound effect of M on FCP behavior is held in sharp contrast with the more moderate influence of M on T_g, yield strength, and elastic modulus [132, 135, 136, 139–141]. We argue that these monotonic properties are controlled by the bulk polymer whereas the fatigue behavior is dominated by the craze zone. In the bulk we would expect to find an assorted array of molecular entanglements containing widely varying chain lengths; consequently, some entanglements should be stronger and more resistant to fracture than another. The monotonic tensile properties should then reflect some weighted average of these assorted entanglements including the relatively weak ones. Therefore, so long as you have a finite number of loosely entangled molecules, the properties of the polymer cannot be dramatically enhanced above a certain critical M and the fracture process will seek out the weaker areas. The situation is different within the crazed material. The crazing process itself has tended to destroy most if not all of the weakly entangled molecular coils. By the very nature of this selection process, only the more highly entangled coils involving the longest molecules remain to form the backbone of the craze fibrils. Consequently, the craze may be envisioned to possess a relatively high M with a relatively narrow molecular weight distribution (MWD). (By contrast, the polymer bulk would have a lower M with a relatively broad MWD.) Since craze stability depends predominantly on the fracture resistance of the fibrils that contain a preponderance of high M entanglements, FCP resistance should show a strong dependence on M since only the highest M fraction is left in the craze to affect its response. (See *Note added in proof*, p. 145.)

The specific nature of the fatigue crack propagation rate dependence on M has been examined further in terms of MWD effects on fatigue behavior in nominally bimodal distribution specimens of PMMA [142]. Samples were prepared to yield high-M and low-M polymer matrices containing varying amounts of low-M, medium-M, and high-M tails. The low-M matrices were designed to have values of M below a critical value of M, M_c, necessary for the development of tensile strength in the glassy polymer [143–145]. In general, the incorporation of medium-M or high-M tails resulted in greater resistance to both static and fatigue loading (Fig. 3.28a). In fact, the existence of high-M tails permitted fatigue fracture testing to be extended into an otherwise "forbidden" range of M_w or M_n. That is, no such tests were possible with control specimens of unimodal MWD when M_w and M_n were less than about 8×10^4 and 4×10^4, respectively [130]. Instead, these specimens broke prior to or during machining. These values are close to those reported for the limiting value of the "zero-strength" M for PMMA, about 3×10^4 [143–145]. These observations are in accordance with the findings of Kambour [134] and Wellinghoff and Baer [138], the latter having noted that small proportions of high-M species can markedly stabilize crazes in polystyrene. Overall, the kinetics of crack propagation were found to depend on one of the higher-average molecular weights such as M_z as opposed to M_n. For example, it was shown that the value of ΔK^* (the value

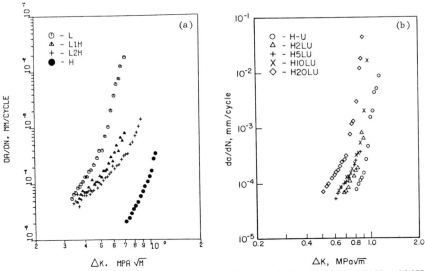

Fig. 3.28 (a) Effect of high-M additions on FCP behavior in low-M PMMA (L1H and L2H correspond to 1 and 2% high-M additions, respectively). [Janiszewski (146)]; (b) effect of low-M additions on FCP behavior in high-M PMMA (Numerals 2,5,10,20 correspond to percentage of low-M added to high-M matrix.) [Kim, Janiszewski, Skibo, Manson, and Hertzberg (142)].

of ΔK required to drive the crack at a given FCP rate) varied directly with M_z for both single and bimodal MWDs (Fig. 3.29). At the other end of the testing spectrum, a low-M tail was found to decrease resistance to FCP (Fig. 3.28b).

The results discussed above support earlier speculations that FCP resistance depends on the stability of crazes ahead of the advancing crack, and in turn on the stability of the entanglement network when subjected to cyclic loading. A spectrum of entanglement stabilities must, of course, exist with the higher-M species probably dominant [143]. Also, during cycling, strain hardening undoubtedly occurs as well, with the higher-M species being most effective in the orientation process involved. Thus, it has been postulated [120] that as the competition between disentanglement and entanglement–orientation proceeds, the load is transferred increasingly to the higher-M species, so that only the most stable (highest-M) craze fibrils are left prior to catastrophic fracture. Preliminary microscopic observations of the fracture surface are consistent with the ability of tails to alter the craze stability and mechanism of failure [146] (see also Section 4.2.2).

In summary, the beneficial effect of a minor addition of a high-M fraction in a low-M matrix polymer is consistent with the fatigue damage model;

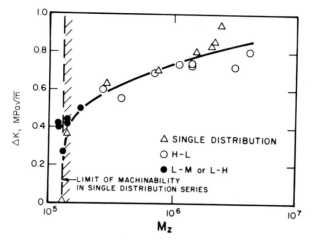

Fig. 3.29   Relationship between ΔK^* (at $da/dN = 10^{-4}$ mm/cycle) and M_z for PMMA: △, single distribution; ○, H–L; ●, L–M or L–H [Kim, Janiszewski, Skibo, Manson, and Hertzberg (142)].

that is, the small number of long chains contained in the high-M tail are largely retained within the craze zone so as to exert a major influence in reducing fatigue crack propagation rates. On the other hand, the addition of low-M species should lead to accelerated FCP rates as a result of entanglement dilution since a low-M tail introduces a disproportionately large number of short molecular chains. Finally, generalizations about the "typical" FCP behavior of a given polymer, including those made previously by the authors [21] (also see Fig. 3.5), must be considered with the reservation that other molecular weights may lead to quite different behavior. Rational prediction of engineering behavior requires specification of M and MWD and laboratory tests conducted under cyclic conditions.

3.8.3   Effect of Cross-Linking on FCP

Cross-linked polymers are finding increased usage in structural members because of their high strength, high modulus and low density; typical applications include composite matrices and adhesive joints. Such materials are not without fault, however, since ductility and toughness properties are often inadequate. In addition, the fatigue response of cross-linked polymers examined to date reveals these materials to exhibit very high crack growth rates at a given ΔK level. The role of cross-linking in hindering segmental flow and, hence, adversely affecting resistance to crack growth is seen in

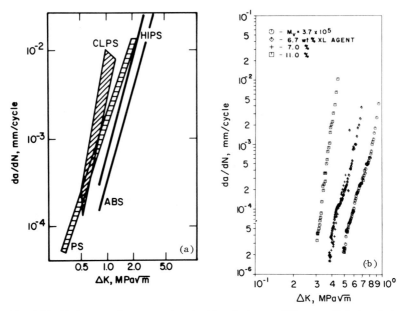

Fig. 3.30 Fatigue crack propagation in (a) polystyrene (PS), cross-linked polystyrene (CLPS), acrylonitrile–butadiene–styrene (ABS), and high impact polystyrene (HIPS) and (b) PMMA with different amounts of cross-linking agents: ○, $M_v = 3.5 \times 10^5$; ◇, 6.7 wt % XL agent; +, 7.0% XL agent; □, 11.0% XL agent. [Reprinted with permission from the American Society for Testing and Materials, 1916 Race St., Philadelphia, PA, 19103 © (40) and courtesy of Quereshi (146a).]

Fig. 3.30a for the case of several styrene-based polymers [21]. The leftward shift of the cross-linked polystyrene FCP results relative to linear polystyrene and the rubber modified blends is consistent with associated changes in fracture energy. One possible explanation for the superior behavior of PS over that of CLPS may involve differences in allowable deformation processes in the two materials; PS is expected to exhibit significantly more crazing than cross-linked polystyrene because of the much tighter network structure in the latter material. To be sure, crazing is a weakening process but, nevertheless, one involving the dissipation of considerable strain energy. It would appear that some mechanism for deformation, however localized, is better than none in making a polymeric material more resistant to static and fatigue fracture.

A similar effect of cross-linking on the fatigue response of PMMA is illustrated by the data shown in Fig. 3.30b. A supply of PMMA ($M_v = 3.7 \times 10^5$) was modified with the addition of a cross-linking agent in amounts ranging from 0.1 to 11% [146a]. No differences in FCP rates were found when the amount of cross-linking agent remained below 3.2%; in sharp contrast, additions of approximately 7 and 11% of the cross-linking agent contributed to a considerable worsening of material resistance to fatigue crack propagation. A similar reduction in fracture toughness was also observed.

Sutton [107] examined the FCP response of a cross-linked commercial epoxy resin and found the FCP rate to vary with ΔK according to Eq. (3.10). Of particular interest, values for the constants A and m were higher than those determined for most other polymeric solids tested to date. More recent studies [147, 148] have been conducted to evaluate the effect of cross-linking density on FCP rates in fully characterized epoxy resins in a manner analogous to the molecular weight studies described in Section 3.8.2. Fatigue crack growth rate data for an amine-cured bisphenol-A-type epoxy are shown in Fig. 3.31 as a function of the amine/epoxy ratio, the latter reflecting differences in the average molecular weight between cross-links, M_c (see Table 3.6). Note the persistent shift to lower fatigue crack growth rates with increasing amine/epoxy ratio (i.e., increasing M_c). In addition, the apparent fracture toughness of these resins showed a steady improvement with increasing M_c. As found earlier by Sutton [107], the exponent m was found to be quite high in these epoxy recipes ($7.7 \leq m \leq 20$). The results of Sutton are included in Fig. 3.31 for comparison.

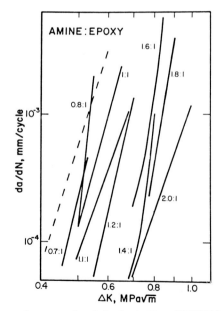

Fig. 3.31 Fatigue crack propagation behavior in Epon 828/MDA epoxy as a function of amine/epoxy ratio at 10 Hz. Dotted line represents data by Sutton [Kim, Skibo, Manson, Hertzberg, and Janiszewski (147)].

TABLE 3.6 Composition and Structure of Epoxy Resins[a]

Amine/epoxy ratio	$M_c{}^b$	Amine/epoxy ratio	$M_c{}^b$
0.7:1	780	1.2:1	415
0.8:1	457	1.4:1	496
0.9:1	314	1.6:1	630
1.0:1	303	1.8:1	896
1.0:1	300	2.0:1	1381
1.1:1	328		

[a] See [147, 148].
[b] Values of M_c were calculated from rubbery moduli derived from dynamic mechanical tests.

As part of this investigation, Kim *et al.* [147, 148] also found that the elastic moduli of these epoxy resins increased generally with increasing amine/epoxy ratio along with the apparent fracture toughness and overall fatigue resistance. In view of the apparent virtues of nonstoichiometric formulation, one might ask: Why not deliberately use strongly nonstoichiometric formulations? The problem is that while several properties are thereby somewhat improved, the values of T_g (at 110 Hz) were found to be decreased from 198 to 119°C as the amine/epoxy (A/E) ratio was increased from 1.0 to 2.2. Consequently, the proper choice of the A/E ratio should depend on a prior establishment of the critical properties that need to be optimized.

We conclude that a correlation exists between molecular chain mobility and fatigue resistance; a similar correlation involving fracture resistance has long been known. Since some plastic deformation at the crack tip is associated with stable fatigue crack propagation, those materials that offer the greatest energy dissipation during deformation (e.g., high M_c resins) should exhibit the most fatigue-resistant behavior. On the other hand, one would expect that the low M_c resins, possessing a tightly constrained structure, would be intolerant of plastic deformation and lead to accelerated fatigue crack growth rates.

3.8.4 Effect of Plasticization on FCP

External plasticizing agents often are added to polymers to enhance processability and make the final polymer product more ductile. At the same time, it should be recognized that the yield strength, elastic modulus, and T_g decrease. Since certain properties improve while others suffer when plasticizers are used, it is appropriate to examine the effect of plasticizer additions on fatigue crack propagation response. Two studies have been conducted which deal with the effect of dioctyl phthalate (DOP) plasticizer additions on the fatigue response of PVC. In one study, Skibo *et al.* [127] found that when 6 and 13% DOP was added to PVC, the FCP rates remained relatively unchanged. It may be noted that these observations are in conflict with those reported by Suzuki *et al.* [149], who reported that da/dN increased strongly as the concentration of DOP was increased from 0 to 10 to 20%. The reason for this discrepancy is not known, though the 40-fold lower cyclic test frequency used by Suzuki *et al.* may have introduced a creep cracking component that would be expected to increase at higher DOP levels.

The effect of DOP on K_c is complex; at a level of 6%, DOP appears to embrittle the polymer [127]. This behavior resembles the so-called "anti-plasticizer" effect, in which small concentrations of plasticizer appear to inhibit segmental mobility and increase modulus and brittleness [150]. On the other hand, at a level of 13%, K_c is reduced only slightly and the effect of DOP on fatigue and fracture properties is minimal.

These observations are quite different from those made in a study of FCP in internally plasticized PMMA [130]. In this case, the use of *n*-butyl acrylate (*n*BA) as an internal plasticizer had a pronounced effect on both FCP and fracture toughness behavior. Property changes tended to fall into two regimes, depending on the *n*BA content. As shown in Fig. 3.32, the effective K_c values and the FCP resistance for the *n*BA–MMA copolymers decreased with increasing *n*BA content up to 20%. When the mole percent of *n*BA was increased to 30, a dramatic apparent increase in toughness occurred along with a reduction in FCP rates. In fact, the 70/30 copolymer behavior became very much like that of a commercial PMMA. (This is not to say that commercial PMMA contains an internal plasticizer—only that the copolymer and commercial materials are similar in terms of FCP resistance and not some other property such as modulus.)

Although the data are limited in number, clearly strong compositional effects are present in some competitive manner. At low *n*BA contents, mechanical crack growth may be dominated by the addition of *n*BA and its effect of increasing ductility at the expense of strength and modulus. The latter effect would be consistent with the tendency of molecules with large effective cross-sectional areas to be weaker than those with smaller ones [151]. Alternately, with a reduction in elastic modulus associated with an increase in *n*Ba content, the growth rate at a given applied ΔK level would be expected to increase [152]. At a higher concentration of *n*BA, overall toughening may occur, this time by crack blunting, rather than by an effect of molecular structure per se. The onset of crack blunting in the 0.3 mole fraction *n*BA copolymer is believed to result from localized crack-tip heating arising from the markedly higher tan δ value for this composition.

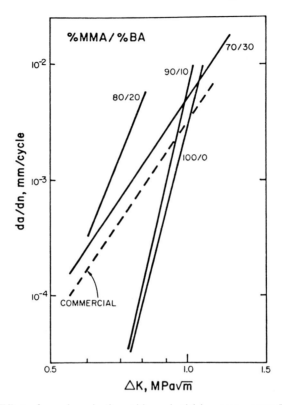

Fig. 3.32 Effect of copolymerization with a plasticizing comonomer BA, on the FCP behavior of PMMA. Conditions: room temperature, 10 Hz [Kim, Skibo, Manson, and Hertzberg (130)].

The observed role played by external and internal plasticizing agents in affecting FCP behavior is consistent with the damage zone model discussed in Sections 3.8.1 and 3.8.2. The addition of the external plasticizing agent DOP in PVC serves to introduce a low-M fraction to the polymer bulk. To be sure, the craze initiation process is made easier (i.e., the craze stress decreases) but the FCP rates remain essentially the same. The latter reflects the fact that the FCP rate is controlled by the fatigue resistance of the craze fibrils that would contain roughly the same amount of high-M fraction as without the DOP addition. (Recall that the model presumes that the crazing process destroys all but the most highly entangled networks involving the longest molecules.) The internal plasticization in PMMA represents a different situation entirely. Here, the addition of nBA into the PMMA chain itself, incorporates the rubbery nBA constituent into the network entanglements that make up the craze fibrils. Therefore, their presence cannot be ignored.

The effect of absorbed moisture on the fatigue behavior of nylon 66 represents an interesting case study involving material plasticization. Effects of moisture on the mechanical behavior of polyamides have long been known, and have been reviewed by McCrum *et al.* [52], Papir *et al.* [153], and Kohan [154]; effects on the ultimate behavior of nylon 66 are also summarized in the trade literature [155]. Thus in general, Young's modulus and yield strength tend to decrease, while impact toughness tends to increase, with increasing water content. Dynamic mechanical spectroscopy and other studies provide evidence that water exists in both tightly bound and weakly bound states [153], and it has been shown that many properties exhibit a transition when water content exceeds the ratio of one water molecule per two amide groups (corresponding to tightly bound water).

Less is known about the effects on fatigue. Fatigue tests on unnotched specimens [155] are reported to indicate that the fatigue strength (the stress corresponding to failure at a given number of cycles) can be reduced by as much as 30% when nylon 66 is equilibrated at 50% relative humidity (RH) (Fig. 3.33). In addition, preliminary fatigue crack propagation (FCP) tests by Manson and Hertzberg [21] showed that "wet" nylon 66 samples that had been soaked in water for six weeks exhibited higher crack growth rates at a given stress intensity factor range than samples of the same polymer which had been vacuum dried. Since the exact moisture levels in these specimens were not measured and the tests were conducted at different cyclic frequencies, a quantitative relationship between water content and FCP behavior was not established.

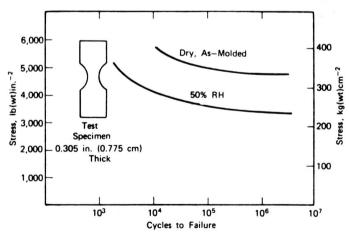

Fig. 3.33 Flexural fatigue life of nylon 66 as-molded and equilibrated to 50% relative humidity. [M. I. Kohan, "Nylon Plastics," © 1973 by John Wiley & Sons, Inc. Reprinted by permission of John Wiley & Sons, Inc.]

More recently, Bretz *et al.* [156, 156a, 156b] determined the specific effect of moisture content on the fatigue crack propagation behavior in nylon 66. Specimens were moisture-conditioned to correspond to equilibration with 0–8.5 wt % water content. The FCP data derived from these samples are shown in Fig. 3.34 at a ΔK level of 3 MPa $\cdot$ m$^{1/2}$ and are plotted as a function of water content. A pronounced minimum in crack growth rate is noted at about 2.5–2.6% water. It is interesting to consider the water/amide group stoichiometry associated with a water content of about 2.6 wt %. Assuming a degree of crystallinity of 40% [156a], and that the water remains in the amorphous phase, 2.6 wt % water would correspond to one water molecule per 3.7 amide groups. In sharp contrast, the FCP rate in the sample containing 8.5% water is 25 times higher. Thus, as water is imbibed by an essentially dry polymer, the FCP rate first decreases to a minimum value and then rapidly increases, reaching a maximum value on saturation. This behavior certainly suggests the existence of competitive effects (and hence mechanisms), the balance depending on the amount of water present.

While a complete rationalization of this interesting behavior must await completion of dynamic mechanical characterization [156b], an explanation must surely take into account the hydrogen-bonding and plasticizing nature of water in polyamides [52, 153, 154]. It has long been known that both the α- and β-transitions are shifted to increasingly lower temperatures by the sorption of increasing proportions of water [52]. However, the rate of de-

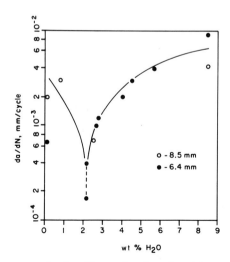

Fig. 3.34 FCP response of nylon 66 as a function of moisture content in two material supplies ($\bigcirc$, 8.5 mm; $\bullet$, 6.4 mm) at $\Delta K = 3$ MPa $\cdot$ m$^{1/2}$ [Bretz (156b)].

crease of T_g is greatest below about 2% water by weight [153, 157, 158]. The decrease in T_g in this range of water content has been attributed to the breaking of hydrogen bonds between amide groups and the formation of water bridges between them [52, 153]. Paradoxically, at the same time that the hydrogen bonds between the polymer molecules are being disrupted, the molecular packing is improved, as shown by a decrease in specific volume of the tightly bound water [159]. At higher concentrations, the water is presumed to be more loosely bound, and presumably serves essentially as a diluent. Thus, the increased mobility of the polyamides at water contents below about 2–3% may well lead to enough localized deformation at the crack tip to permit blunting of the crack, thus decreasing the proportion of strain energy available for crack propagation, and lowering the FCP rate at a given ΔK. On saturation, however, gross plasticization undoubtedly occurs, as is evident in a 2.5-fold reduction in the room temperature Young's modulus E [154], in comparison with the values for samples equilibrated at up to 50% RH ($\sim 2.6\%$ water overall). Such dilution and reduced intermolecular attraction between polymer molecules may be expected to result in a decrease in the constraint (lower E) exerted on the crack tip by the bulk material ahead of it, and hence an *increase* in the FCP rate at a given value of ΔK. In other words, the strain per loading cycle $\Delta \varepsilon$ must then increase, since $\Delta \varepsilon = \Delta \sigma / E$. The saturated nylon must then accumulate more damage per loading cycle than would be the case with the drier specimens; a higher FCP rate will then be expected at a given value of $\Delta \sigma$ (and hence ΔK).

Additional evidence for enhanced chain mobility in samples containing > 2.5 wt % water was observed by noting a rise in temperature at the crack tip during the fatigue test. Whereas the dry specimen and the one equilibrated at 0.8 wt % experienced no noticeable temperature rise during testing, the crack-tip temperature of the higher moisture content specimens rose an estimated 8–20°C. Such a temperature change is indicative of an irreversible deformation process such as large-scale chain motion at the crack tip. Thus if crack-tip heating is localized, permitting localized deformation and crack blunting, FCP rates may be decreased; however, if the heating involves more than a small volume element at the crack tip, the consequent softening is believed to result in enhanced specimen compliance and higher crack growth rates. The latter condition has been encountered in the case of impact modified nylon 66 [80, 80a] (see Section 5.1.5).

The relationship of these results to the question of fatigue strength and endurance must now be considered. Clearly, with these tests the longest fatigue life may be expected for specimens containing 2–3% water (corresponding to equilibration at about 50% RH). However, an apparent contradiction between the present findings and the previously mentioned results by Kohan [154] requires resolution. The latter results were based on S–N plots (i.e., maximum stress versus number of cycles to failure). It is known that nylon 66 experiences a significant temperature elevation during S–N tests because of its relatively large loss tangent. This temperature elevation is amplified if the polymer contains moisture (note observations above); therefore, it would be expected that nylon 66 equilibrated at 50% RH would experience a greater temperature rise during a fatigue test than a dry sample. This temperature elevation, in turn, would result in a decrease in the modulus of nylon, the decrease being greater for the 50% RH sample than for the sample in the as-molded condition (see Section 2.3.2). Since the specimens were fatigued between fixed load limits, the 50% RH sample would experience greater cyclic strains and, therefore, more damage per cycle than the dry specimens. For this reason, the nylon 66 samples equilibrated at 50% RH would be expected to exhibit inferior fatigue properties in the S–N test. In the present FCP tests, on the other hand, hysteretic heating is not sufficient to cause a drop in modulus at this level of water content, because the cyclic deformation is confined to the crack-tip region with the bulk of the specimen acting as heat sink, whereas in the S–N tests, the whole specimen experiences the load. As discussed previously, the effect of moisture in this test is to allow for increased plastic deformation and crack-tip blunting, thereby lowering crack growth rates. Hence, these FCP tests, conducted on *notched* rather than unnotched specimens, leads to the conclusion that nylon 66 equilibrated with 2.5 wt % water ($\sim 50\%$ RH) is the superior material under cyclic loading conditions. By comparison, the inferior FCP response of the fully saturated nylon 66 samples can be rationalized in terms of the much lower elastic modulus throughout the specimen, thus allowing for large amounts of cyclic damage within the fixed load limits of the crack growth experiment. Therefore, the relative fatigue resistance of a nylon 66 component as a function of moisture content is found to vary with the test procedure, in this case whether one uses notched or unnotched specimens. This fact must be clearly recognized when making component design decisions.

3.8.5 Effect of Crystallinity on FCP

It is interesting to note that the polymers most resistant to crack propagation appear to be crystalline: The lowest fatigue crack growth rates at a given ΔK level have been recorded in nylon 66 [117, 129, 156], nylon 66 toughened blends [80], nylon 6 [40], poly(vinylidiene fluoride) (PVDF) [40], and polyacetal (PA) [129, 160] (Fig. 3.35). An even more dramatic comparison of data that confirms this finding is found in Fig. 3.36. The FCP rates of an amorphous polyamide (Amidel) are shown to be approximately three

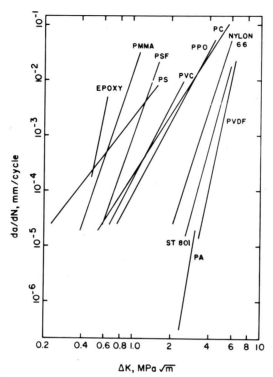

Fig. 3.35 Fatigue data for several polymeric solids. Note superior fatigue resistance in crystalline polymers (Nylon 66, ST801, PA, PVDF). Epoxy ($A/E = 1.0$), PMMA ($M_v = 1.25 \times 10^6$), PSF ($M_v = 5 \times 10^4$), PS ($M_v = 2.7 \times 10^5$), PVC ($M_w = 2.25 \times 10^5$), PPO, PC ($M_v = 4.9 \times 10^4$), Nylon 66 ($M_w = 17,000$, 2.2% H_2O), ST801 (0% H_2O), PA ($M_n = 7 \times 10^4$), PVDF ($M_w = 2.2 \times 10^5$).

orders-of-magnitude greater than those associated with semicrystalline nylon 66. To be sure, the chemistry of these two polyamide materials is somewhat different. However, the differences in FCP rate are so great that crystallinity undoubtedly plays a major if not dominant role. There are some apparent exceptions: chlorinated polyether and low-density polyethylene, both of which exhibit intermediate fatigue resistance [40]. However, normalization of the polyethylene data to allow for its very low modulus shows that its comparative ranking is actually very high (see Section 3.8.7). In the case of chlorinated polyether, its performance is still similar to that of, say, high-impact polystyrene.

It seems likely that the superior fatigue resistance of crystalline polymers relative to amorphous structures is not accidental. As pointed out by Koo *et al.* [161, 162] and by Meinel and Peterlin [163], crystalline polymers not only can dissipate energy when crystallites are deformed, but they can also

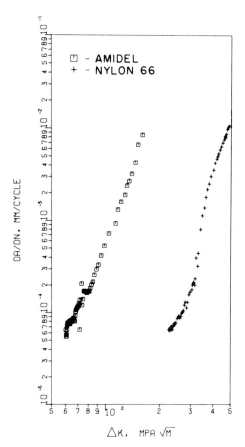

Fig. 3.36 Comparison of fatigue crack growth rates in semicrystalline nylon 66 (Zytel 101, +) and amorphous nylon (Amidel, □).

apparently reform a crystalline structure that is exceedingly strong. As such, the fatigue fracture process must involve some considerable amount of cold drawing.

The toughening effect of crystallinity is not confined to plastics. Strain-crystallizable rubbers are superior to noncrystallizing ones. Indeed, Andrews [90] has pointed out a possible analogy to a particular case of a strain-crystallizable rubber. On cycling natural rubber, crystallization occurs on straining but melting occurs on relaxation. If, however, the cycle does not pass through zero strain, the crystalline material cannot melt, and the growth per cycle is significantly reduced. Thus, a permanently crystalline polymer might be expected to behave in a similar fashion. This point was discussed in Section 3.7 and illustrated by Fig. 3.23.

In light of the markedly superior FCP response of crystalline polymers, it is ironic that almost nothing has been published concerning the effect of spherulite size and percent crystallinity on fatigue behavior. In one of the few revealing studies conducted to date, Laghouati *et al.* [131, 164] found a fourfold decrease in FCP rates in high-density polyethylene when the percent crystallinity increased from 47 to 55 and no significant effect of spherulite size on fatigue response was noted. For further discussion of the effect of spherulite size on fatigue resistance see Section 3.9.

3.8.6 Effect of Thermal History on FCP

It is well recognized that thermal history (e.g., annealing treatments and cooling rates) has a strong influence on free volume, excess free volume, and degree of crystallinity which, in turn, affect the mechanical properties of polymeric solids. Unfortunately, few studies have been conducted to evaluate the role played by thermal history in controlling the fatigue behavior of these materials. In the study just mentioned, Laghouati *et al.* [131, 164] observed that FCP rates in high-density polyethylene decreased as the cooling time from the molding temperature to 100°C decreased from 8 to 0.5 min (Fig. 3.37). They concluded that a higher concentration of tie molecules are

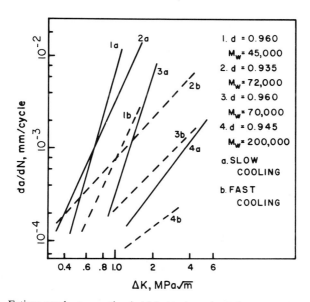

Fig. 3.37 Fatigue crack propagation in high density polyethylene as function of molecular weight and cooling rate. [de Charentenay, Laghouati, and Dewas (131)].

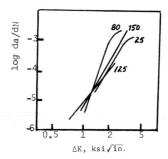

Fig. 3.38 Fatigue crack propagation response in polycarbonate as a function of annealing temperature (°C) [Manson, Hertzberg, Kim, and Wu (56)].

likely to exist in the intercrystalline regions of the rapidly quenched samples and that these tie molecules would enhance the crack growth resistance of the material.

Manson *et al.* [56] obtained preliminary FCP measurements on PC that had been annealed at 80, 125, and 150°C. It had been expected that annealing below T_g (150°C) would result in closer packing within the amorphous matrix while annealing at T_g would induce crystallization [165]. They found that annealing at 80°C shifted the growth curve to a lower ΔK range while the other treatments had little effect (Fig. 3.38). Though these thermal treatments did not influence fatigue properties greatly, a strong correlation was noted between the effective fracture toughness ΔK_f (defined as the last ΔK value recorded prior to rapid fracture) and the magnitude of the endothermal peak Q_a, introduced by the annealing treatment. Figure 3.39 shows that when Q_a increased, the apparent fracture toughness decreased. A reasonable explanation for this relationship may be proposed. The parameter Q_a should in some manner reflect the degree of ordering in PC. If so, one may hypothesize a competition between the ordering or close packing of small segments, facilitated by increasing annealing temperature, and disordering caused by

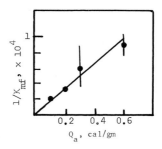

Fig. 3.39 Relationship between ΔK_{max} and Q_a for annealed polycarbonate [Manson, Hertzberg, Kim, and Wu (56)].

the onset of large segmental motions as T_g is approached. (Note that Q_a was greatest at an annealing temperature of 120°C while ΔK_f was lowest at this same temperature.) Thus as ordering increases and free volume is decreased, the ability to relax within the time scale of a stress cycle and dissipate energy through plastic deformation is reduced.

3.8.7 Normalization Parameters for FCP Data

One may conclude from the preceding discussions, that empirical modeling of fatigue crack propagation data in engineering plastics in terms of fracture mechanics concepts has been quite successful. To be sure, FCP response of different polymers depends strongly on both external test variables such as cyclic frequency, temperature, environment, and stress state and also on material variables reflecting compositional and architectural variables. A logical question then arises: Can one normalize these results in a manner that could simplify the correlation of data? One approach involves normalization of ΔK values with respect to Young's modulus E. This procedure has proven successful in the normalization of FCP rates among different metallic alloys but does not significantly reduce the range of ΔK within which stable fatigue crack growth occurs in polymers [152, 160]. For example, the elastic moduli of PMMA and PC as determined from a tensile test are not too dissimilar and yet the FCP rates of these two materials differ by approximately three orders of magnitude. It is interesting to note, though, that on a $\Delta K/E$ basis, fatigue crack growth rates in crystalline polymers are lower than in metallic alloys (Fig. 3.40) (also recall Fig. 3.10b). One major uncertainty regarding the normalization of polymer da/dN versus ΔK data involves the proper choice of elastic constant. Not only does the complex modulus contain storage and loss components, but the experimental values obtained from tensile tests and dynamic measurements differ by as much as a factor of two [41].

A striking correlation of polymer fatigue data is evident, however, between values of ΔK^* (the value of ΔK required to drive a crack at a constant value of da/dN) and ΔK_{max} (the maximum value of ΔK associated with stable fatigue crack growth) [21, 166] (Fig. 3.41). Since ΔK_{max} represents some measure of fracture toughness K_c, it is seen that the greater the toughness of a polymer, the greater the driving force required to drive a crack at constant speed. Both Martin and Gerberich [85] and Mostovoy and Ripling [167] also reported a generally well-behaved relationship between ΔK^* and ΔK_{max}, and Mostovoy and Ripling reported an essentially equivalent relationship between da/dN at constant ΔK and ΔK_{max}. Thus, as suggested in Wnuk's equation (3.14), the crack growth rate is inversely dependent on the material

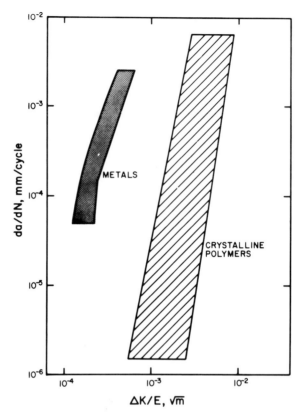

Fig. 3.40 Comparison of crack growth rates in metals and crystalline polymers as a function of normalized stress intensity factor $\Delta K/E$. [Reprinted from R. W. Hertzberg, M. D. Skibo, and J. A. Manson, *J. Mater. Sci.* **13**, 1978, 1038, Chapman and Hall, Ltd.].

toughness. Intuitively, this implies that the mechanisms controlling unstable and catastrophic fracture also control stable fatigue crack extension. Quantitatively, an envelope with values of 0.67 and 0.5 for upper and lower bounds to $\Delta K^*/\Delta K_{max}$ includes data for most polymers. Of course, within this envelope, "fine structure" may well exist, and "best" slopes may differ from one polymer system to another.

From a phenomenological point of view, this finding is consistent with frequent observations that a variety of large-deformation inelastic processes in polymers are closely related to small-deformation (linearly elastic) characteristics such as dynamic mechanical response. For example, a number of mechanical properties such as impact strength, yield strength, fracture energy, and time-to-failure can be correlated in various ways with damping

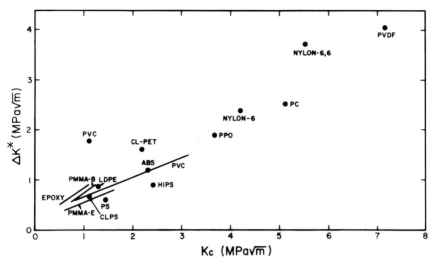

Fig. 3.41 Relationship between ΔK^* (at $da/dN = 7.5 \times 10^{-4}$ mm/cycle) and K_c, a measure of polymer fracture toughness corresponding to the maximum stress intensity range at failure.

behavior, often with the occurrence of β-transitions at low temperatures. The correlation between FCP frequency sensitivity and the β-transition is illustrative of this point (recall Section 3.4).

3.9 MATERIAL SELECTION AND COMPONENT LIFE PREDICTION

   The fatigue crack propagation resistance of a particular polymer relative to another can be determined directly by examination of data such as shown in Fig. 3.5, 3.30–3.32, and 3.35–3.38. For example, we see from Fig. 3.5 that the FCP resistance of nylon 66 and PVDF is far superior to that of, say, PMMA and PS. In addition, the maximum stress intensity level attainable with these two crystalline polymers is much greater than that associated with the glassy PMMA and PS glassy materials (recall Fig. 3.41). This means that if identically cracked components of polystyrene and nylon 66 were subjected to the same stress fluctuation, the crystalline nylon 66 part would endure a much larger number of cycles prior to failure *and* allow for the development of a much larger flaw size before the onset of unstable fracture. Moreover, crack growth rate relationships, whether derived from theoretical considerations or based on empirical data, may be used to compute such things as the component life and the effective size of the original flaw. For example, if Eqs. (3.7) and (3.10) are combined, the da/dN (ΔK) relationship assumes the form

$$da/dN = A\,\Delta K^m = A Y^m\,\Delta\sigma^m\,a^{m/2}. \tag{3.21}$$

In turn, Eq. (3.21) can be integrated to yield (for $m \neq 2$) when Y does not change within the limits of integration

$$N_f = \frac{2}{(m-2)A Y^m\,\Delta\sigma^m}\left(\frac{1}{a_0^{[m-2/2]}} - \frac{1}{a_f^{[m-2/2]}}\right), \tag{3.22}$$

where N_f represents the number of cycles to failure, a_0 is the initial crack size, a_f the final crack size at failure, $\Delta\sigma$ the stress range, A, m are material constants from fatigue crack propagation test results, and Y is a stress intensity calibration factor. Note that the computation of cyclic life depends on knowledge of such variables as the applied stress range, the stress intensity calibration factor Y, the material constants A and m and the starting and final flaw sizes. In practical situations involving a small initial flaw size and a large final crack dimension, the computed component life is found to depend markedly on the initial defect size but not on the final flaw size.

From the above, it is important for one to obtain an accurate measure of the initial flaw size in order to perform a component life computation of this type. Unfortunately, such information is sometimes difficult to obtain because of limitations in nondestructive inspection procedures. However, it is possible to compute an effective initial flaw using Eq. (3.22). For example, if the total life of the sample or component is known along with the applied stress range, material variables (A and m), correction factor Y, and the final flaw size at failure, then Eq. (3.22) or a similar relationship based on Eq. (3.12) may be used to derive an initial effective flaw size. Naturally, this presumes that no loading cycles are required to initiate the starting flaw. Supporting evidence for this analysis procedure has been reported for the fatigue fracture of polyethylene [168] and cellular polyurethane elastomers [169]. In the former case, the computed size of the inferred starting flaw was numerically equal to the crystalline spherulite diameter. In related fashion, the defect size required to initiate fatigue failure in a cellular elastomer was found to be equal in size to the inherent largest pore dimension in the cellular polyurethane material.

Such agreement between computed starting flaw sizes and actual microstructural dimensions not only lends confidence to the computational method, but also demonstrates the importance of the propagation stage of fatigue fracture in predicting overall component life. In addition, by identifying through computational methods a specific microstructural component that serves as the initiation site for the fatigue process, it is then possible to focus attention on seeking one or more thermomechanical treatments that reduce the scale of the suspect microstructural feature. For example, the polyethylene study showed the starting flaw size to be comparable to the spherulite diameter. Consequently, fatigue life should improve with decreasing spherulite size. This may be a contributing factor to account for the improved FCP resistance of impact-modified nylon 66 relative to nylon 66 [80]. Also, Bareishis and Stinskas [170] reported superior cyclic life in polycaproamide when nucleating agents were used to produce a significant decrease in spherulite size.

Of major practical importance, Eq. (3.22) may be used to predict the service life of an engineering component. In this instance, the initial flaw size is either confirmed by direct examination, inferred on the basis of nondestructive test information, or estimated based on prior field experience. [The final flaw size a_f can be computed from Eq. (3.8) for the crack instability condition where $K_c = Y\sigma a_f^{1/2}$.] If the computed life for the given set of service conditions is less than the desired service life for the component, the part would have to be replaced or the service stresses reduced to some safer level. On the other hand, if the computed fatigue life safely exceeds the desired service life of the product, the cracked part could be left in service. There are obvious economic benefits associated with the latter situation since unnecessary component replacements could be avoided.

Two hypothetical design problems are now considered to illustrate the fatigue life computational procedure [171]. In the first case, a large flat plate is to be fabricated from any one of several polymers. If the plate were to

TABLE 3.7 Computed Fatigue Lives of Plastic Plates[a]

Material	Final crack length (cm)	Fatigue life (no. of cycles)
Polystyrene ($M = 2.7 \times 10^5$)[b]	7.25	20,200
ABS	35.5	95,900
Acrylic ($M = 4.8 \times 10^6$)	8.2	266,700
PVC ($M = 2.25 \times 10^5$)	24.6	402,500
Polycarbonate ($M \approx 4.8 \times 10^4$)	39.8	1,031,000
Nylon 6/6	38.4	2,902,000
Poly(vinylidene fluoride)	44.8	9,826,000

[a] See [171].
[b] M = molecular weight.

experience a cyclic stress of 3 MPa and contain an initial through-thickness crack of length 2.5 cm, how long would it be expected to endure? Assuming that all of the potential materials exhibit fatigue behavior conforming to the FCP relationship given by Eq. (3.10), it is first necessary to establish the limits of integration by determining the critical crack length where unstable cracking occurs. Table 3.7 reveals the critical crack lengths for several polymeric solids along with the number of fatigue cycles ($\Delta\sigma$ = 3 MPa) necessary to grow the crack in each plate from 2.5 cm to the critical size for that material. Note that the tolerable flaw size varies dramatically from one engineering plastic to another, reflecting considerable differences in fracture toughness. Also, the cyclic lives prior to failure are vastly different among the materials examined in the exercise.

The second materials selection design problem involves a 10-mm-thick large diameter liquid storage cylinder, which is to be fabricated with any one of several engineering plastics. Assuming that the cylinders contain a semi-circular surface crack 2-mm-deep and oriented normal to a fluctuating stress of 15 MPa, how many loading cycles would be required for the crack to grow completely through the cylinder wall?

The results of these computations are shown in Fig. 3.42. Again we see a

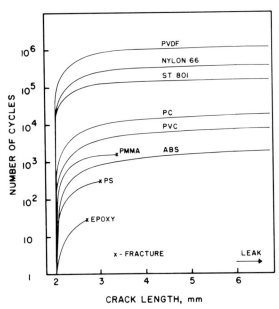

Fig. 3.42 Computed fatigue lives of plastic storage tanks based on hypothetical conditions described in text and fatigue data generated at 10 Hz [Hertzberg and Manson (171)].

substantial difference in computed fatigue lives among the materials examined. Of particular note, tanks made from acrylic and PS would never leak prior to their catastrophic failure. Rather, conditions for unstable crack growth would be met in these two materials prior to the time when the crack would have grown through the cylinder wall. This latter condition should be avoided if at all possible since the "warning stage" of impending danger (i.e., leaking) would be precluded by abrupt fracture. Additional examples of fatigue life prediction are given in Section 5.3.1.

REFERENCES

[1] R. W. Hertzberg, "Deformation and Fracture Mechanics of Engineering Materials." Wiley, New York, 1976.

[2] A. A. Griffith, *Trans. R. Soc. London* 221 (1920).

[3] C. E. Inglis, *Proc. Inst. Naval Architects* p. 60 (1913).

[4] E. Orowan, *Phys. Soc. Rep. Prog. Phys.* **12,** 186 (1948).

[5] J. P. Berry, "Fracture Processes in Polymeric Solids" (B. Rosen, ed.). Wiley (Interscience), New York, 1964.

[6] F. A. Johnson and J. C. Radon, *Eng. Fract. Mech.* **4,** 555 (1972).

[7] J. C. Radon, *Polym. Eng. Sci.* **12,** 425 (1972).

[8] H. K. Mueller and W. G. Knauss, *Trans. Soc. Rheol.* **15,** 217 (1971).

[8a] S. J. Bennett, G. P. Anderson and M. L. Williams, *J. Appl. Polym. Sci.,* **14,** 735 (1970).

[9] G. R. Irwin, Fracture, Handbuch der Physik (S. Flugge, ed.), Vol. VI, p. 551 (1958).

[10] H. M. Westergaard, *J. Appl. Mech. Trans. ASME* (June 1939).

[11] G. R. Irwin, *Trans. ASME Ser. D* **82**(2), 417 (1960).

[12] F. A. McClintock and G. R. Irwin, ASTM STP **381,** 84 (1965).

[13] Annual Book of ASTM Standards, ASTM Standard E399-78 (1979).

[14] P. C. Paris and G. C. M. Sih, ASTM STP 381, p. 30 (1965).

[15] H. Tada, P. C. Paris and G. R. Irwin, "The Stress Analysis of Cracks Handbook." Del Research Corporation, St. Louis, Missouri, 1973.

[16] G. C. M. Sih, "Handbook of Stress Intensity Factors." Lehigh Univ., 1973.

[17] H. F. Bueckner, *J. Appl. Mech. Trans. ASME,* **80,** 1225 (1958).

[18] P. C. Paris, *Proc. Sagamore Army Mater. Res. Conf., 10th* p. 107 (1964).

[19] P. C. Paris and F. Erdogan, *J. Bas. Eng. Trans. ASME Ser. D* **85**(4), 528 (1963).

[20] P. C. Paris, Ph.D. Dissertation, Lehigh Univ. (September 1962).

[21] J. A. Manson and R. W. Hertzberg, *CRC Crit. Rev. Macromol. Sci.* **1**(4), p. 433 (1973).

[22] R. S. Rivlin and A. G. Thomas, *J. Polym. Sci.* **10,** 291 (1953).

[23] G. J. Lake, P. B. Lindley, and A. G. Thomas, *Proc. Int. Conf. Fracture, 2nd, Brighton* p. 493. Chapman and Hall, London, 1969.

[24] G. J. Lake and P. B. Lindley, *Conf. Ser. No. 1* p. 176. Institute of Physics and Physics Society, London, 1966.

[25] G. J. Lake and P. B. Lindley, *J. Appl. Polym. Sci.* **8,** 707 (1964).

[26] A. N. Gent, P. B. Lindley, and A. G. Thomas, *J. Appl. Polym. Sci.* **8,** 455 (1964).

[27] A. G. Thomas, Conference Series No. 1, Inst. Phys. Physical Soc., Oxford (September 1966).

[28] A. G. Thomas, *J. Polym. Sci.* **31,** 467 (1958).

[29] N. E. Waters, *J. Mater. Sci.* **1**, 354 (1966).
[30] H. F. Borduas, L. E. Culver, and D. J. Burns, *J. Strain Anal.* **3**, 193 (1968).
[31] N. H. Watts and D. J. Burns, *Polym. Eng. Sci.* **7**, 90 (1967).
[32] S. Arad, J. C. Radon, and L. E. Culver, *J. Mech. Eng. Sci.* **13**, 75 (1971).
[33] B. Mukherjee and D. J. Burns, *Exp. Mech.* **11**, 433 (1971).
[34] R. W. Hertzberg, J. A. Manson, and M. D. Skibo, *Polym. Eng. Sci.* **15**, 252 (1975).
[35] M. D. Skibo, R. W. Hertzberg, and J. A. Manson, *J. Mater. Sci.* **11**, 479 (1976).
[36] R. W. Hertzberg and J. A. Manson, *J. Mater. Sci.* **8**, 1554 (1973).
[37] J. C. Radon, *J. Appl. Polym. Sci.* **17**, 3515 (1973).
[38] J. P. Elinck, J. C. Bauwens, and G. Homes, *Int. J. Fract. Mech.* **7**(3), 227 (1971).
[39] P. G. Faulkner and J. R. Atkinson, *J. Appl. Polym. Sci.* **15**, 209 (1971).
[40] R. W. Hertzberg, J. A. Manson, and W. C. Wu, ASTM STP 536, p. 391 (1973).
[41] R. W. Hertzberg, M. D. Skibo, J. A. Manson, and J. K. Donald, *J. Mater. Sci.* **14**, 1754 (1979).
[42] S. Arad, J. C. Radon, and L. E. Culver, *J. Appl. Polym. Sci.* **17**, 1467 (1973).
[43] S. Arad, J. C. Radon, and L. E. Culver, *Eng. Fract. Mech.* **4**, 511 (1972).
[44] H. M. El-Hakeem, Ph.D. Thesis, London Univ. (1975).
[45] M. D. Skibo, R. W. Hertzberg, and J. A. Manson, *J. Mater. Sci.* **12**, 531 (1977).
[46] W. G. Knauss and H. Dietmann, *Int. J. Eng. Sci.* **8**, 643 (1970).
[47] M. P. Wnuk, *J. Fract. Mech.* **7**, 217 (1971).
[48] M. P. Wnuk, Prog. Rep., NASA Grant NGR 42-003-006 (May 1971).
[49] M. P. Wnuk, *J. Appl. Mech.* **41**(1), 234 (1974).
[50] J. G. Williams, *J. Mater. Sci.* **12**, 2525 (1977).
[51] J. Koppelman, *Rheol. Acta* **1**, 20 (1958).
[52] N. G. McCrum, B. E. Read, and G. Williams, "Anelastic and Dielectric Effects in Polymeric Solids." Wiley, New York, 1967.
[53] R. N. Haward, "The Physics of Glassy Polymers" (R. N. Haward, ed.), p. 41. Halstead Press, New York, 1973.
[54] E. G. Bobalek and R. M. Evans, *SPE Trans.* **1**, 93 (1961).
[55] M. D. Skibo, Ph.D. Dissertation, Lehigh Univ. (1977).
[56] J. A. Manson, R. W. Hertzberg, S. L. Kim, and W. C. Wu, "Toughness and Brittleness of Plastics" (R. D. Deanin and A. M. Crugnola, eds.), p. 146. American Chemical Society, New York, 1976.
[57] J. S. Harris and I. M. Ward, *J. Mater. Sci.* **8**, 1655 (1973).
[58] R. W. Hertzberg, *Closed Loop* **3**(6), 12 (1973).
[59] S. Arad, J. C. Radon, and L. E. Culver, *Polym. Eng. Sci.* **12**, 193 (1972).
[60] M. D. Skibo, unpublished research.
[61] P. Beardmore and S. Rabinowitz, "Treatise on Materials Science and Technology" (R. J. Arsenault, ed.), Vol. 6, p. 267. Academic Press, New York, 1975.
[62] K. Oberbach, *Kunststoffe, Bd.* **63**, 37 (1973).
[63] J. A. Manson, R. W. Hertzberg, S. L. Kim, and M. D. Skibo, *Polymer* **16**, 850 (1975).
[64] M. D. Skibo, R. W. Hertzberg, and J. A. Manson, Fracture 1977, Vol. 3, ICF4, p. 1127. Waterloo, Canada, 1977.
[65] R. W. Hertzberg, J. A. Manson, and M. D. Skibo, *Polymer* **19**, 359 (1978).
[66] S. de Petris, V. Frosini, E. Butta, and M. Baccarreda, *Makromol. Chem.* **109**, 54 (1967).
[67] J. Stoelting, F. E. Karasz, and W. J. Macknight, *Polym. Eng. Sci.* **10**, 133 (1970).
[68] A. Eisenberg and B. Cayrol, *J. Polym. Sci. Polym. Symp.* **35**, 129 (1971).
[69] E. Butta, S. de Petris, and M. Pasquini, *Ric. Sci.* **38**, 927 (1968).
[70] H. Oberst, *Kunstoffe* **53**, 1 (1963).
[71] J. Heijboer, *J. Polym. Sci. Polym. Symp.* **16**, 3755 (1968).

[72] J. Heijboer, *Br. Polym. J.* **1**, 3 (1969).

[73] R. F. Boyer, *Polym. Eng. Sci.* **8**, 161 (1968).

[74] P. I. Vincent, *Polymer* **15**, 111 (1974).

[75] W. Retting, *Eur. Polym. J.* **6**, 853 (1970).

[76] J. C. Bauwens, *J. Polym. Sci. Polym. Symp.* **33**, 123 (1971).

[77] L. J. Broutman and T. Kobayashi, *Int. Conf. Dyn. Crack Propagation, Lehigh Univ.* (1972).

[78] J. C. Radon and L. E. Culver, *Polym. Eng. Sci.* **15**(7), 500 (1975).

[79] R. Attermo and G. Ostberg, *Int. J. Fract. Mech.* **7**, 122 (1971).

[80] M. D. Skibo, R. W. Hertzberg, and J. A. Manson, "Deformation, Yield and Fracture of Polymers," p. 4.1. Plastics and Rubber Institute, Cambridge, 1979.

[80a] R. W. Hertzberg, M. D. Skibo and J. A. Manson, ASTM STP 700, (1980).

[80b] E. A. Flexman, Jr., "Toughening of Plastics," p. 14.1. Plastics and Rubber Institute, London, 1978.

[81] M. L. Williams, R. F. Landel, and J. D. Ferry, *J. Am. Chem. Soc.* **77**, 3701 (1955).

[82] T. Kurobe and H. Wakashima, *Jpn. Congr. Mater. Res.—Non-Metall. Mater.*, *13th* 192 (1970).

[83] T. Kurobe and H. Wakashima, *Jpn. Congr. Mater. Res.—Non-Metall. Mater.*, *15th* 137 (March 1972).

[84] Y. M. Mai and J. G. Williams, *J. Mater. Sci.* **14**(8), 1933 (1979).

[85] G. C. Martin and W. W. Gerberich, *J. Mater. Sci.* **11**, 231 (1976).

[86] R. J. Wann, G. C. Martin, and W. W. Gerberich, *Polym. Eng. Sci.* **16**(9), 645 (1976).

[87] W. W. Gerberich and G. C. Martin, *J. Polym. Sci. Polym. Phys. Ed.* **14**, 897 (1976).

[88] W. W. Gerberich, private communication.

[89] G. P. Marshall and J. G. Williams, *Int. Conf. Corros. Fatigue*. Nat. Assoc. of Corros. Engineers and Am. Inst. of Metallurgical Engineers, Storrs, Connecticut (January 14–18, 1971).

[90] E. H. Andrews, "Fracture of Polymers." American Elsevier, New York, 1968.

[91] G. J. Lake and P. B. Lindley, *J. Appl. Polym. Sci.* **9**, 1233 (1965).

[92] G. J. Lake and P. B. Lindley, *J. Appl. Polym. Sci.* **9**, 2031 (1965).

[93] M. Braden and A. N. Gent, *J. Appl. Polym. Sci.* **3**, 90 (1960).

[94] G. P. Marshall, L. E. Culver, and J. G. Williams, *Proc. R. Soc. London Ser. A* **319**, 165 (1970).

[95] G. P. Marshall, L. E. Culver, and J. G. Williams, *Plast. Polym.* **38**, 95 (1970).

[96] P. G. Faulkner and J. R. Atkinson, *Plast. Polym.* **143**, 109 (1972).

[97] J. G. Williams and G. P. Marshall, "Deformation and Fracture of High Polymers" (H. H. Kausch, J. A. Hassel, and R. I. Jaffe, eds.), p. 557. Plenum Press, New York, 1973.

[97a] G. P. Marshall and J. G. Williams, *Int. Conf. Corros. Fatigue, Univ. Connecticut* p. 38/2 (1971).

[98] Y. W. Mai, *J. Mater. Sci.* **11**, 303 (1976).

[99] H. G. Krenz, Relationships Between Structure and Micromechanics of Solvent Crazes in Glassy Polymers. Ph.D. Dissertation, Cornell Univ. (1977).

[100] H. A. El-Hakeem and L. E. Culver, *J. Appl. Polym. Sci.* **22**, 2691 (1978).

[101] J. R. Martin and J. F. Johnson, *J. Appl. Polym. Sci.* **18**, 3227 (1974).

[102] Y. W. Mai, *J. Mater. Sci.* **9**, 1896 (1974).

[103] S. Arad, J. C. Radon, and L. E. Culver, *J. Mech. Eng. Sci.* **14**(5), 328 (1972).

[104] R. Roberts and F. Erdogan, *J. Bas. Eng. Trans. ASME* **89**, 885 (1967).

[105] R. G. Forman, V. E. Kearney, and R. M. Engle, *J. Bas. Eng. Trans. ASME* **89**, 459 (1967).

[106] J. A. Sauer, A. D. McMaster, and D. R. Morrow, *J. Macromol. Sci.-Phys.* **B12**(4), 535 (1976).

[107] S. A. Sutton, *Eng. Fract. Mech.*, **6**, 587 (1974).

[108] N. J. Mills and N. Walker, *Polymer* **17**(4), 335 (1976).

[109] K. Yamada and M. Suzuki, *Kobunshi Kagaku* **30**(336), 206 (1973).

[110] S. M. Cadwell, R. A. Merrill, C. M. Sloman, and F. L. Yost, *Ind. Eng. Chem. Anal. Ed.* **12**(1), 19 (1940).

[111] A. R. Bunsell and J. W. S. Hearle, *J. Mater. Sci.* **6**, 1303 (1971).

[112] E. H. Andrews, *J. Appl. Phys.* **32**(3), 542 (1961).

[113] E. H. Andrews, *J. Mater. Sci.* **9**, 887 (1974).

[114] F. J. Pitoniak, A. F. Grandt, L. T. Montulli, and P. F. Packman, *Eng. Fract. Mech.* **6**, 663 (1974).

[115] F. J. Pitoniak, AFML-TR-72-235 (November 1972).

[116] D. H. Banasiak, A. F. Grandt, Jr., and L. T. Montulli, *J. Appl. Polym. Sci.* **21**, 1297 (1977).

[116a] Y. W. Mai, *Int. J. Fract.* **15**, 103 (1979).

[117] R. W. Hertzberg, H. Nordberg, and J. A. Manson, *J. Mater. Sci.* **5**, 521 (1970).

[118] N. J. Mills, "Deformation, Yield and Fracture of Polymers," p. 6.1. Plastics and Rubber Institute, Cambridge, 1973.

[119] H. R. Brown and I. M. Ward, *Polymer* **14**, 469 (1973).

[120] R. W. Hertzberg, M. D. Skibo, and J. A. Manson, ASTM STP 675, 471 (1979).

[121] D. S. Dugdale, *J. Mech. Phys. Solids* **8**, 100 (1960).

[122] N. I. Muskhelishvili, "Some Basic Problems of the Mathematical Theory of Elasticity." Noordhoff, Groningen, Holland, 1953.

[123] E. Foden, D. R. Morrow, and J. A. Sauer, *J. Appl. Polym. Sci.* **16**, 519 (1972).

[124] J. A. Sauer, E. Foden, and D. R. Morrow, *Polym. Eng. Sci.* **17**, 246 (1977).

[125] J. A. Sauer, *Polymer* **19**, 859 (1978).

[126] S. Warty, J. A. Sauer, and A. Charlesby, *Eur. Polym. J.*, **15**(5), 445 (1979).

[127] M. D. Skibo, J. A. Manson and R. W. Hertzberg, *J. Macromol. Sci.-Phys.* **B14**(4), 525 (1977).

[128] C. Rimnac *et al.* (manuscript in preparation).

[129] P. E. Bretz *et al.* (manuscript in preparation).

[130] S. L. Kim, M. D. Skibo, J. A. Manson, and R. W. Hertzberg, *Polym. Eng. Sci.* **17**(3), 194 (1977).

[131] F. X. de Charentenay, F. Laghouati, and J. Dewas, "Deformation, Yield and Fracture of Polymers," p. 6.1. Plastics and Rubber Institute, Cambridge, 1979.

[132] J. P. Berry, *J. Polym. Sci. Polym. Lett. Ed.* 4069 (1964).

[133] H. H. Kausch, *Kunstoffe* **65**, 1 (1976).

[134] R. P. Kambour, *J. Polym. Sci. Macromol. Rev.* **7**, p. 1 (1973).

[135] R. P. Kusy and D. T. Turner, *Polymer* **17**, 161 (1976).

[136] R. P. Kusy and D. T. Turner, *Polymer* **15**, 394 (1974).

[137] G. W. Weidmann and W. Döll, *Colloid Polym. Sci.* **254**, 205 (1976).

[138] S. Wellinghoff and E. Baer, *J. Macromol. Sci.-Phys.* **B11**, 367 (1976).

[139] J. R. Martin, J. F. Johnson, and A. R. Cooper, *J. Macromol. Sci.-Rev. Macromol. Chem.* **C8** 57 (1972).

[140] E. H. Merz, L. E. Nielsen, and R. Buchdahl, *Ind. Eng. Chem.* **43**, 1396 (1951).

[141] H. W. McCormick, F. M. Brower, and L. Kin, *J. Polym. Sci.* **39**, 87 (1959).

[142] S. L. Kim, J. Janiszewski, M. D. Skibo, J. A. Manson, and R. W. Hertzberg, *ACS Organic Coatings and Plast. Chem.* **38**(1), 317 (1978).

[143] J. D. Ferry, "Viscoelastic Properties of Polymers," 2nd ed. Wiley, New York, 1970.

[144] G. C. Berry and T. G. Fox, *Adv. Polym. Sci.* **5**, 261 (1968).

[145] A. N. Gent and A. G. Thomas, *J. Polym. Sci. Polym. Lett. Ed.* **10**, p. 571 (1972).

[146] J. Janiszewski, M. S. Thesis, Lehigh Univ. (1978).

[146a] J. Quereshi, Private communication.

[147] S. L. Kim, M. D. Skibo, J. A. Manson, R. W. Hertzberg, and J. Janiszewski, *Polym. Eng. Sci.* **18**(14), 1093 (1978).

[148] J. A. Manson, L. H. Sperling and S. L. Kim, Final Rep. AFML-TR-124 (May 1975–April 1977).

[149] K. Suzuki, S. Yada, N. Mabuchi, K. Seiuchi, and Y. Matsutani, *Kobunshi Kagaku*, **28**, 920 (1971).

[150] P. I. Vincent, *Polymer* **1**, 425 (1960).

[151] P. Vincent, *Polymer* **13**, 558 (1972).

[152] S. Pearson, *Nature (London)* **211**, 1077 (1966).

[153] Y. S. Papir, S. Kapur, C. E. Rogers, and E. Baer, *J. Polym. Sci. Polym. Lett. Ed.* **10**, 1305 (1972).

[154] M. I. Kohan, "Nylon Plastics." Wiley, New York, 1973.

[155] "Zytel" Design Handbook. E. I. duPont de Nemours and Co., Wilmington, Delaware (1972).

[156] P. E. Bretz, R. W. Hertzberg, and J. A. Manson, *J. Mater. Sci.* **14**, 2482 (1979).

[156a] P. E. Bretz, R. W. Hertzberg, J. A. Manson and A. Ramirez, ACS Symposium Series, 1980, in press.

[156b] P. E. Bretz, Ph.D. Dissertation, Lehigh Univ. (1980).

[157] D. C. Prevorsek, R. H. Butler, and H. Reimschussel, *J. Polym. Sci. Polym. Lett. Ed.* **9**, 867 (1971).

[158] K. H. Illers, *Makromol. Chem.* **38**, 168 (1960).

[159] H. W. Starkweather, *J. Macromol. Sci.* **B3**, p. 727 (1969).

[160] R. W. Hertzberg, M. D. Skibo, and J. A. Manson, *J. Mater. Sci.* **13**, 1038 (1978).

[161] G. P. Koo, "Fluoropolymers, High Polymers" (L. A. Wall, ed.), Vol. XXV, p. 507. Wiley (Interscience), New York, 1972.

[162] G. P. Koo and L. G. Roldan, *J. Polym. Sci. Polym. Lett. Ed.* **10**, 1145 (1972).

[163] G. Meinel and A. Peterlin, *J. Polym. Sci. Polym. Lett. Ed.* **9**, 67 (1971).

[164] A. F. Laghouati, Ph.D. Dissertation, 3rd Cycle, Univ. de Technologie de Compiegne, 1977.

[165] D. G. LeGrand, *J. Appl. Polym. Sci.* **13**, 2129 (1969).

[166] J. A. Manson and R. W. Hertzberg (manuscript in preparation).

[167] S. Mostovoy and E. Ripling, *Polym. Sci. Technol.*, **9B** (Adhes. Sci. Technol., L. H. Lee, ed.), 513 (1975).

[168] E. H. Andrews and B. J. Walker, *Proc. R. Soc. London Ser. A* **325**, 57 (1971).

[169] R. E. Whittaker, *J. Appl. Polym. Sci.* **18**, 2339 (1974).

[170] I. P. Bareishis and A. V. Stinskas, *Mekh. Polim.* **9**(3), 562 (1973).

[171] R. W. Hertzberg and J. A. Manson, *Plast. World* **35**(5), 50 (1977).

[172] G. Pitman and I. M. Ward, *J. Mater. Sci.* **15**, 635 (1980).

Note added in proof: Pitman and Ward [172] have expanded studies of the effect of M on FCP behavior to include polycarbonate. As found previously for other polymers (Figs. 3.25–3.27), the FCP rate in PC at a given ΔK level decreased markedly with increasing M_n and M_w.

4 | *Fatigue Fracture Micromechanisms in Engineering Plastics*

It is well recognized that careful examination of fine-scale fracture surface details can provide significant information regarding various fracture processes in solids [1, 2]. Often such studies have revealed some microstructural defect(s) that were responsible for the component's demise. On the basis of such enlightening observations, materials engineers are better able to improve overall component response through changes in composition and internal microstructure of the material. In addition, fractographic investigations often provide useful quantitative information that can be used to analyze the continuum details of the fracture process. For these reasons, numerous failure analyses involving metal alloys have been reported, which make considerable use of fractographic information [3, 4]. As such, it will be helpful to examine polymer fatigue fracture surface markings in this chapter within the context of the considerable literature for metals.

4.1 MACROFRACTOGRAPHY OF FATIGUE FAILURES

Many distinctive fracture surface markings are readily apparent from a macroscopic examination of a component that has been subjected to repeated loadings. More often than not, the fatigue fracture surface is oriented perpendicular to the principal stress direction. A casual examination of such surfaces very often reveals a series of concentric lines (usually curved) that radiate from one or more sources within the component body or at its surface (Fig. 4.1). These arrest lines, often called "clam shell markings" or

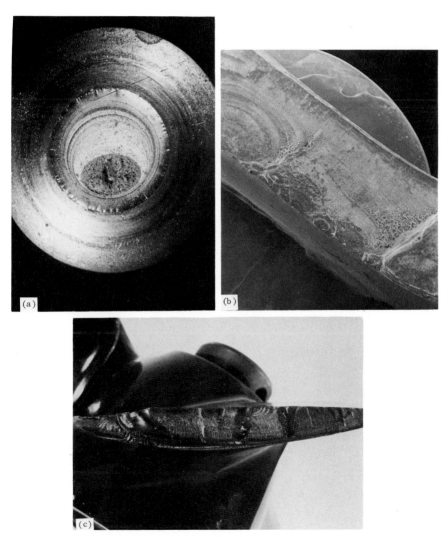

Fig. 4.1 Typical fatigue "clam shell markings" in (a) steel alloy, (b) plastic tricycle, (c) marine engine plastic propeller blade.

"beach markings" have been attributed to different periods of crack extension during the life of the component. It is important to recognize that these concentric bands represent *periods* of growth and are not representative of *individual* load excursions. For the case of engineering solids, such beach markings may arise from such conditions as intermittent changes in stress amplitude, variable test frequency and environment, and changing mean stress level. The center of curvature of each set of beach markings represents the fatigue origin. Hence, these fracture bands are extremely useful in that they direct the investigator to the vulnerable region(s) in the component that served as crack nucleation sites.

When a crack progresses across a relatively narrow section of a component, shear lips may form at the free surfaces. That is, the plane of the crack assumes a $\pm 45°$ angle relative to the stress axis and component thickness. It has been shown that the shear lips reflect crack propagation under conditions of plane stress and the flat (i.e., $90°$ orientation) region corresponds to plane strain conditions [4]. For the case of high-strength steels and aluminum alloys, it has been possible to relate the width of the shear lip to the plane stress plastic zone radius (Fig. 4.2a). Hence, a measurement of the shear lip width can lead to an estimate of the prevailing stress intensity conditions [see Eq. (4.1)]

$$D \approx r_y \approx \frac{1}{2\pi}\left(\frac{K}{\sigma_{ys}}\right)^2. \tag{4.1}$$

Since $K = Y\sigma a^{1/2}$, Eq. (4.1) may be rewritten as

$$D \approx \frac{1}{2\pi}\frac{Y^2\sigma^2 a}{\sigma_{ys}^2}. \tag{4.2}$$

Therefore, if one knows the K calibration factor Y from Eq. (3.8), the crack length a where the shear lip was measured, and the material's yield strength σ_{ys}, Eq. (4.2) can be used to estimate the stress level associated with a specific shear lip width.

More recently, this analytical procedure has been extended to the analysis of polymer fatigue fracture surfaces. Hertzberg and Manson [5] have observed shear lip formation during FCP tests of polycarbonate (Fig. 4.2b). As expected, the width of the shear lips increased with increasing crack length, consistent with an increasing stress intensity factor. By setting the plane stress crack-tip plastic zone size r_y to be approximately equal to the shear lip width [4], a measure of material yield strength could be estimated from Eq. (4.2). Using shear lip measurements at different crack length positions and relating these to the corresponding stress intensity factor, the material's yield strength was computed and found to be in rea-

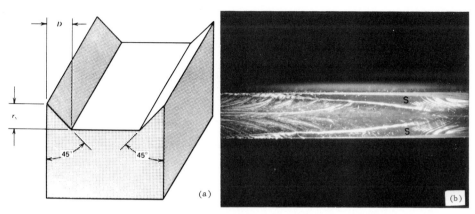

Fig. 4.2 Shear lip development on fracture surface: (a) Relationship between shear lip depth *D* and estimated plane stress plastic zone size [R. W. Hertzberg, "Deformation and Fracture Mechanics of Engineering Materials," © 1976 by John Wiley & Sons, Inc. Reprinted by permission of John Wiley & Sons, Inc.] and (b) shear lip formation (*S*) during fatigue crack growth in polycarbonate, 2.67 × [Reprinted with permission from *Crit. Rev. Macromol. Sci.* **1**(4), 1973, 433. Copyright The Chemical Rubber Co., CRC Press Inc.]

sonable agreement with reported values. The fact that most polymers do not exhibit classical shear lips may be related to the propensity of these materials toward craze-controlled yielding on planes normal to the stress axis. (See *Note added in proof*, p. 183.)

Some useful information regarding the fatigue process in polymers also can be obtained by noting changes in the color and texture of polymer fatigue fracture surfaces. For example, a number of amorphous polymers, such as polycarbonate (PC), polysulfone (PSF), polystyrene (PS), and low-molecular-weight poly(methyl methacrylate) (PMMA), develop a mirror-smooth, transparent fatigue fracture surface at low ΔK levels that transforms to a rougher texture along with a misty appearance at intermediate and high ΔK values [6] (Fig. 4.3). Beginning at the mirror-mist transition region and continuing for some distance beyond, a brilliant array of packets of color fringes is found on these fracture surfaces. Within the mirror region, only one color is revealed for each material. Berry [7] and Kambour [8, 9] reported previously that the existence of such fringes reflects the presence of craze material. The relatively uniform color of the mirror region probably reflects the existence of single craze development (see Section 4.2.3) and the color fringe packets in the mist region reflect craze bundling (i.e., multiple parallel crazes). It should be recognized, however, that craze-induced color fringes are not restricted to fatigue fracture surfaces and can be detected on surfaces generated under monotonic loading conditions [7–9].

In FCP studies of semicrystalline polymers including various polyamides,

Fig. 4.3 Mirror and mist fatigue fracture surface appearance in (a) poly(vinyl chloride), (b) polycarbonate, (c) polysulfone, (d) polystyrene, (e) poly(methyl methacrylate) [M. D. Skibo, R. W. Hertzberg, J. A. Manson, and S. Kim, *J. Mater. Sci.* **12**, 1977, 531, Chapman and Hall, Ltd.].

polyacetal, chlorinated polyether, and poly(vinylidene fluoride) (PVDF), stress whitening has been observed on the surface of the sample (Fig. 4.4a) and on the fracture surface (Fig. 4.4b). Such whitening is believed to be associated with refractive index changes occurring as a result of molecular chain reorientation and spherulite breakdown in the slowly moving crack tip plastic zone (see Section 3.8.5). It has been found that the size of the damaged region increases with increasing crack length in accord with an associated increase in the crack tip plastic zone size [Eq. (4.1)] [10]. Furthermore, it has been shown that whitening occurs in the region of stable fatigue crack advance and in the plastic zone existing ahead of the crack at the point of crack instability. Beyond this damage zone no further whitening is observed when the crack moves abruptly across the remaining unbroken ligament. Overall, the fatigued surfaces are relatively rough and occasionally reflect passage of the crack either around or through the spherulites [11, 12].

Rubber-modified amorphous polymers also reveal stress whitening in the slow-stable crack growth regime. In these materials, stress whitening, which increases with increasing ΔK levels, is believed to be caused by multi-

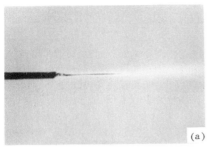

(a)   (b)

Fig. 4.4 Stress whitening during fatigue of semicrystalline polymers. (a) Side view in PVDF; (b) Fracture surface in chlorinated polyether. Stress whitening occurs only during fatigue loading [Reprinted with permission from ASTM STP 536, 1973, 391. Copyright ASTM, 1916 Race St., Philadelphia, PA 19103.]

ple crazing in association with second phase rubber particles. Again, the amount of whitening decreases abruptly at the point of fast fracture.

4.2 MICROFRACTOGRAPHY OF FATIGUE FAILURES

From the previous discussion, we saw that it was possible to extract quantitative information from some fatigue fracture surface markings. It will now be shown that other microscopic markings provide even more useful and reliable information for purposes of failure analysis and provide a clearer understanding of polymeric fatigue crack propagation processes. While some of these additional fracture surface features can be studied with light optics, others require the application of electron microscope techniques.

Before discussing detailed observations resulting from an electron microscopic examination, it is necessary to recognize some of the unique problems encountered in the microscopy of polymeric solids. When using the transmission electron microscope, laboratory technicians almost invariably have used a cellulose acetate replicating tape to prepare their metallic samples for examination in the microscope. This material is first softened with acetone before being pressed onto the fracture surface. It must be recognized that acetone acts as an aggressive solvent and softener of most polymeric solids. Hence, the application of the acetone-softened tape would begin immediately to destroy the fracture surface itself. Alternately, poly(acrylic acid), which is soluble in water, was found to be useful as an alternate replicating media. Even so, the investigator is faced with the problem of artifacts being generated when removing the replica from the sample. Since many fracture surfaces of polymeric solids contain a layer of crazed material, mechanically stripping the replicating material from the sample tends to disturb the fracture surface and introduces artifacts onto the replicating film. Subsequent replications merely serve to compound the problem. For example, it has been shown that some fracture surface details, found in transmission electron micrographs, were not found in samples examined with the scanning electron microscope [13]. Furthermore, samples that were initially examined with the scanning electron microscope showed different details than samples that were replicated prior to examination in the scanning microscope. These observations strongly suggest that the replication procedure can have an adverse effect on the integrity of the fracture surface. Consequently, it is suggested by the authors that most fractographic studies of polymeric solids should be conducted with the scanning electron microscope, unless it can be proven that replicating procedures in a given material do not introduce spurious microdetails.

The use of the scanning electron microscope is not without its own diffi-

Fig. 4.5 SEM beam damage in high-M PMMA with 10% low-M addition $\Delta K = 0.91$ MPa·m$^{1/2}$ [Janiszewski, Hertzberg and Manson (55)].

culties, however. First, the sample must be coated with either carbon or a heavy metal so as to provide for efficient charge transfer. Even so, the electron beam tends to destroy the fracture surface topology either by localized melting or breakdown in the spherulitic structure of crystalline solids (Fig. 4.5). For those materials that do deteriorate under the electron beam, the fractographic examinations must be conducted at a lower accelerating voltage. Unfortunately, this reduces the resolution of the instrument. One is therefore faced with a dilemma in that higher accelerating voltages will provide better resolution but lead to accelerated deterioration of the fracture surface.

For many of the fracture surfaces to be discussed in this chapter, the following laboratory procedures were utilized. Specimens prepared for use in the SEM were coated first with carbon and then with gold. The accelerating potential was set at 20 kV for all materials except PMMA and polyacetal (PA), which were examined at 5 kV. Replicas for TEM study were made by replicating the fracture surface with 10% aqueous poly(acrylic acid). After drying, the replica was chromium shadowed and carbon coated. The plastic replica was then dissolved in warm water, leaving a carbon replica to be mounted on a grid for viewing.

Whenever possible, attempts were made not only to describe the fracture micromorphology but also to determine the quantitative relationships that might exist between the fracture markings and the prevailing stress conditions. To this end, fracture surface studies were conducted with samples that had been subjected to carefully controlled fatigue crack propagation experiments. In this regard, the macroscopic crack growth rate for each sample was monitored and related to the prevailing stress intensity factor condition. As a result, it was possible to determine the fracture surface morphology under known stress conditions.

4.2.1 Striation Formation

From the extant metal fatigue literature, one finds abundant evidence for the existence of fatigue striations. These markings (Fig. 4.6a) correspond to the successive positions of the advancing crack front as a result of individual load excursions; hence the spacing between each line represents the incremental crack advance during one load excursion. Therefore, the striations not only identify a cyclic loading condition, but also provide quantitative information regarding the kinetics of the fatigue cracking process. Such markings have been reported in a vast number of different metal alloy systems with their width being found to vary with the prevailing stress intensity conditions at the crack tip. A useful empirical relationship [14]

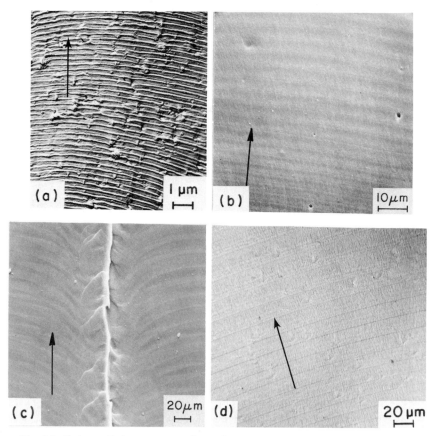

Fig. 4.6 Fatigue striations corresponding to incremental advance of a crack as a result of a single loading cycle. Arrows indicate crack growth direction. (a) Aluminum alloy, (b) polycarbonate, (c) polysulfone, and (d) poly(methyl methacrylate) [M. D. Skibo, R. W. Hertzberg, J. A. Manson and S. Kim, *J. Mater. Sci.* **12**, 1977, 531, Chapman and Hall, Ltd.].

between the striation spacing and the stress intensity factor is given by

$$\text{striation spacing} \approx 6(\Delta K/E)^2, \tag{4.3}$$

where ΔK is the stress intensity factor range $Y \Delta \sigma \, a^{1/2}$ and E the modulus of elasticity. Since the stress intensity factor is a function of the stress σ and defect size a, Eq. (4.3) can be used to infer the prevailing stress level at any location on the fracture surface where a reliable striation width measurement can be made. In a sense then, fatigue striations serve as a permanent record of the fracture process. It has been suggested that this second-power dependence of striation width on ΔK reflects the fact that the striation

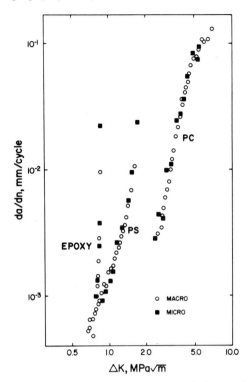

Fig. 4.7 Comparison of macroscopic growth rates and striation spacing measurements in epoxy, polystyrene, and polycarbonate: ○, macro; ■, micro [Reprinted with permission from ASTM STP 675, 1979. Copyright ASTM, 1916 Race St., Philadelphia, PA, 19103 (37).]

width represents a fraction of the crack opening displacement [15]. It is important to note that the macroscopic growth rate for metals varies with ΔK raised to the power 2.5–7 or more. This implies that the macroscopic growth rate constitutes a summation of several fracture mechanisms that may be described by

$$\frac{da}{dN_{\text{macro}}} = \sum Af(K)_{\substack{\text{striation} \\ \text{mechanism}}} + Bf'(K)_{\substack{\text{void} \\ \text{coalescence}}}$$

$$+ Cf''(K)_{\text{cleavage}} + Df'''(K)_{\substack{\text{corrosion} \\ \text{component}}} + \cdots. \qquad (4.4)$$

Consequently, one finds ΔK regimes wherein striations are either larger or smaller than the average value of the macroscopic growth rate.

It is appropriate, therefore, to determine whether polymeric solids generate similar fatigue markings on fracture surfaces produced under cyclic loading conditions. The earliest reported studies of polymer fatigue

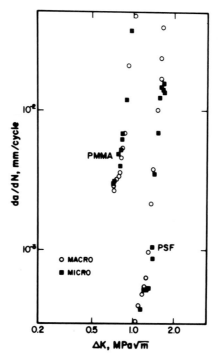

Fig. 4.8 Comparison of macroscopic growth rates and striation spacing measurements in polysulfone and poly(methyl methacrylate): ○, macro; ■, micro [Reprinted with permission from ASTM STP 675, 1979. Copyright ASTM, 1916 Race St., Philadelphia, PA, 19103 (37).]

fracture surfaces, using optical microscopes did, in fact, record the presence of fatigue striations in rubber and in such commercial engineering plastics as poly(methyl methacrylate) and polycarbonate [16–23]. Furthermore, striation width was found to increase with increasing crack length, consistent with the tendency for the striation spacing to increase with increasing stress intensity factor (see Eq. 4.3). More recently, it was shown that striation spacings in PMMA and rubber increased with a power of ΔK greater than two [24, 25]. More extensive quantitative data are now available to determine whether a relationship similar to that given by Eq. (4.3) exists in polymeric solids (i.e., whether polymer striation spacings vary with ΔK^2 and whether other mechanisms in addition to striations are found at a given ΔK level). Figures 4.7 and 4.8 show both macroscopic and microscopic test results for PMMA, polysulfone, polystyrene, polycarbonate, and an epoxy resin. In each case, one finds an excellent one-to-one agreement between striation width and macroscopic growth rate. This implies that in the ΔK regime where striations are found, they represent the only fatigue crack advance micromechanism and account for the entire crack growth increment during a given loading cycle. By way of confirmation, no other fracture mechanism is seen in these materials in the ΔK regime where striations are found. Also in sharp contrast with observations from metal fracture surfaces, it is obvious that the ΔK dependence of polymer striation width is not constant (and equal to two) but rather varies with the power of ΔK ranging from 4 to about 20! Obviously, striation width–crack opening displacement correlations are suspect for the case of polymer fatigue. Furthermore, post-failure analytical procedures aimed at using striation data to infer the prevailing ΔK conditions [e.g., Eq. (4.3) for the case of metals] are not possible at this time since a parallel unifying relationship for polymers has not been defined. Instead, investigators must have previous knowledge of a particular polymer's FCP behavior (e.g., da/dN versus ΔK data) before using striation spacing width measurements to infer ΔK conditions.

Representative fractographs showing fatigue striations in several polymers are shown in Fig. 4.6. These fracture surface markings are generally flat, especially at lower ΔK levels. Recently, Kitagawa [26] made use of interference microscope techniques to show that the striation width-to-depth ratio in PMMA was relatively constant at $100:1$ (Fig. 4.9).

From Fig. 4.6b–d, we see that the surface of each striation contains a fine linear structure oriented normal to the striation line itself. Such lines may reflect evidence of excessive material tearing during the striation formation process. Overall, there appears to be no change in morphology

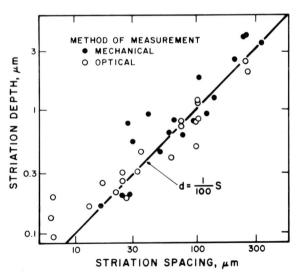

Fig. 4.9 Relationship between striation depth and spacing. Method of measurement: ●, mechanical; ○, optical [Kitagawa (26)].

from one side of the striation to the other, in marked contrast to the discontinuous growth bands to be discussed in Section 4.2.2. The slight striation curvature is similar to that found in metal samples and indicates the crack propagation direction (see arrows in Fig. 4.6).

An understanding of fatigue fracture surface micromorphology in crystalline polymers has received limited attention to date. It is known that crack advance occurs faster in certain spherulites than in others as evidenced by the development of an uneven crack front [27] (Fig. 4.10a). This is completely analogous to the nature of fatigue crack advance in polycrystalline metallic solids (Fig. 4.10b).

White and Teh [27a], Tomkins and Biggs [28], and Bretz [10] have reported striationlike markings parallel to the advancing crack front in nylon, PE, and PVDF. At large ΔK levels near final fracture, Bretz [10] found an excellent one-to-one correlation between striation width and the associated macroscopic growth rate in nylon 66, similar to that shown in Figs. 4.7 and 4.8, while White and Teh matched striation widths with macroscopic growth rates in LDPE. At lower ΔK levels in these materials, however, the fatigue fracture markings become irregular in spacing and take on a different character. For example, a series of orthogonal lines may be seen on the fatigue fracture surfaces of several semicrystalline polymers (Fig. 4.11). Since the crack propagation direction is from left to right in each photograph, the vertical lines could be interpreted as being fatigue striations. Such is not the case, however, as shown by the lack of correlation between the fracture line widths and the associated macroscopic crack growth rates. For example, White and Teh [27a] found that the "microstriation" band widths in LDPE

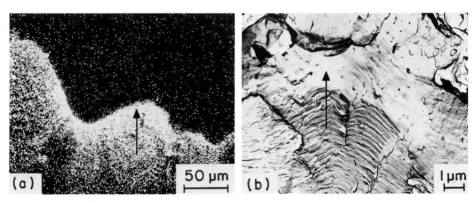

Fig. 4.10 Irregular crack front contour in (a) low density polyethylene as revealed by Al-marked fracture surface (Al x-ray image) [V. I. Singian, J. W. Teh, and J. R. White, *J. Mater. Sci.* **11**, 1976, 703, Chapman and Hall, Ltd.] and (b) aluminum alloy. Arrows indicate crack growth direction.

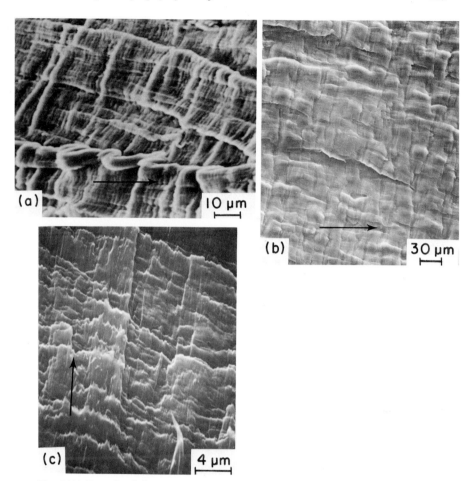

Fig. 4.11 Irregular fatigue markings in crystalline polymers. (a) LDPE [Reproduced from J. R. White and J. W. Teh, *Polymer*, 1979, **20**, 764, by permission of the publishers, IPC Business Press Ltd. ©]; (b) Nylon 66; and (c) PVDF [Bretz (10)]. Note fracture lines oriented parallel and perpendicular to crack direction (arrows).

remained unchanged across the specimen (approximately 0.5 μm), fully an order of magnitude smaller than the measured da/dN value for this region. These investigators have suggested instead that the regions between these vertical lines represent fractured sections or exposed ends of individual lamellae that become aligned as a result of large-scale crack tip plastic strains. Since crystalline lamellae are usually found to be only a few hundredths of a micron thick, however, we believe that the 0.5-μm-wide bands found in Fig. 4.11a may represent instead a cluster of several lamellae or different sections through these crystalline units. White and Teh have suggested that the ragged horizontal lines on the LDPE fatigue fracture surface may represent interspherulitic crack paths at slightly different levels. That is, the distance between these horizontal lines should correspond to the spherulite diameter. Further studies are in progress to clarify the more complex nature of crystalline polymer fatigue fracture surface micromorphology. In this regard, specific attention is being given toward establishing correlations between fracture surface markings and microstructural features.

The morphology and mechanism(s) of fatigue striation formation have been considered by several investigators [17, 20, 21, 23, 29]. McEvily *et al.* [20] and Feltner [17] described striation formation in PMMA, PC, and polyethylene (PE) in terms of the crack-tip blunting model proposed by Laird and Smith [30], wherein the crack tip is considered to blunt and resharpen during loading and unloading portions of the stress cycle, respectively. This model predicts that each fracture surface would consist of flat regions separated by matching troughs on the two surfaces. While McEvily *et al.* [20] observed this to be the case in PE, Feltner [17] noted in PMMA that ". . . the crack profile is roughly saw tooth, with the teeth on one side of the crack fitting into notches on the other side." As mentioned earlier, Kitagawa [26] demonstrated that the striation width-to-depth ratio in PMMA was relatively constant at 100:1. To further complicate the picture, occasional serrated striations have been observed on some fatigue surfaces in PMMA [17, 29] but not in others [19]. Johnson has concluded that the serrated striation will occur in low molecular weight but not in high-molecular-weight PMMA. He further postulated that these serrations might be attributed to local variations in the density of the material through which the crack is propagating. The possible significance of these serrated "striations" will be considered further in Section 4.2.2.

In concluding this section, it is emphasized that with prior determination of the macroscopic fatigue crack growth, it was possible to establish clearly that the markings discussed in this section are striations. This point is reinforced in the following discussion.

4.2.2 Discontinuous Crack Growth

Reflecting the existence of considerable evidence of the one-load-cycle–one-fracture-band relationship in metals and plastics, it has become commonplace for investigators instinctively to conclude that a series of parallel fracture bands oriented normal to the crack direction represented individual striations. Often, this assumption was not verified since neither the macroscopic crack growth kinetics were monitored nor the micromorphology of the bands examined. As a result, investigators have been misled since, in some polymers under certain test conditions, fatigue cracks had progressed across the sample in discontinuous increments associated with a large number of loading cycles [25, 31]. In connection with these observations, a second type of fatigue arrest lines was identified initially in PVC [32, 33] and PC [34] but they did not correspond to the increment of crack growth resulting from a single load excursion. Instead, these bands reflected discrete crack advance increments that occurred after several hundred loading cycles of total crack arrest (Fig. 4.12). For the case of commercial PVC, discontinuous crack growth constituted the only observed mode of crack advance prior to final specimen rupture; striations were never observed on the smooth, mirrorlike PVC fracture surfaces. On the other hand, both striations and discontinuous growth bands were found on PC failure fracture surfaces [34].

These preliminary findings have posed a major challenge to the materials engineer/scientist. For one thing, how does one know whether a given set of fatigue fracture markings indicate crack advance from individual loading cycles rather than increments of hundreds of cycles? At least for cases involving laboratory specimens, the significance of the fracture bands could be

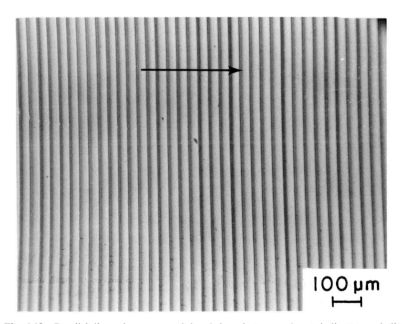

Fig. 4.12 Parallel discontinuous growth bands in polystyrene. Arrow indicates crack direction [M. D. Skibo, R. W. Hertzberg, and J. A. Manson, *J. Mater. Sci.* **11**, 1976, 479, Chapman and Hall, Ltd.].

determined by comparing their width with monitored macroscopic growth rates. As such, one can not emphasize too strongly the desirability of generating companion macroscopic fatigue crack growth rate data when careful fatigue fracture surface micromorphological studies are planned. Obviously such data are not available in actual service failures. Therefore, it has become necessary to identify clearly defined differences between the macro- and micromorphology of striations and the larger discontinuous crack growth bands. Some useful information in this regard can be obtained from optical microscope studies. For example, it is generally found, at least in homogeneous amorphous polymers, that the discontinuous crack growth bands are restricted to the mirror region of the fracture surface. This correlation holds both in precracked and unnotched fatigue test samples [6, 35]. By contrast, fatigue striations are not visible until the crack has grown well into the mist zone. In this connection, the region of the fracture surface where strong color fringes are found, i.e., in the mirror–mist transition region (see Section 4.1), serves to separate these two major sets of fatigue markings. Therefore, discontinuous growth bands would be expected at shorter crack lengths prior to the development of color fringes and striations would be expected at longer crack lengths after dissipation of the interference fringes. Also, for the case of these homogeneous amorphous polymers, the mirror region where discontinuous growth bands form tends to be extremely flat with few if any tear ridges running parallel to the direction of crack growth. For this reason, it is not unusual to observe discontinuous growth bands extending from one surface of the sample to the other without any perturbation. By contrast, the mist zone where striations are found is typically heavily ridged (except for the case of high molecular weight PMMA at high ΔK levels). This results in the development of numerous parallel packets of striations that change curvature abruptly at each ridge (Fig. 4.6b).

The distinction between striations and discontinuous growth bands in crystalline and two-phase polymers is more difficult to identify since one does not find on these fracture surfaces a mirror region or a zone containing interference color fringes. However, the tendency for the development of unperturbed discontinuous growth bands across the specimen width still persists, while the striations tend to form in small packets as previously noted.

Attention is now given to additional quantitative characteristics of the discontinuous growth bands. In a manner distinct from fatigue striation width–stress intensity factor relationships, Elinck *et al.* [32] performed calculations to show that the band width varied with the square of the prevailing stress intensity conditions and corresponded to the dimension of the crack-tip plastic zone as defined by the Dugdale–Muskelishvili formulation [36]. [See Eq. (4.5).] It was found subsequently that a craze zone grew continuously with load cycling, though characterized by a decreasing rate with increasing craze length. When some critical condition was satisfied, the crack would

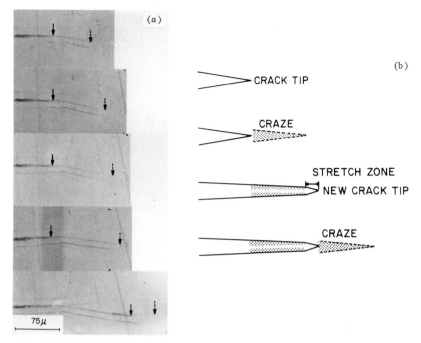

Fig. 4.13 Discontinuous crack growth process. (a) Composite micrograph of PVC showing position of craze (↓) and crack (↓) tip at given cyclic intervals; (b) model of discontinuous cracking process [R. W. Hertzberg and J. A. Manson, *J. Mater. Sci.* **8**, 1973, 1554, Chapman and Hall, Ltd.].

suddenly strike through the entire craze before arresting at the craze tip [29]. In fact, audible clicks have been associated with these discontinuous cracking events. The photographic collage shown in Fig. 4.13a documents this discontinuous cracking process. The arrows in each photograph represent the crack tip and craze-tip locations, respectively, after several blocks of loading cycles (N). Note that the crack tip remains fixed until more than 300 load cycles were applied beyond the starting reference point. The sequence involving continuous craze growth and discontinuous crack growth is modeled in Fig. 4.13b.

Based on recent cinematographic studies [22], the craze grows to about 80% of its final equilibrium length l_D, the Dugdale plastic zone width at that particular K_{max} value, within the first 10% of the band's cyclic life (Fig. 4.14). Little growth then occurs for a substantial portion of the total craze life. Finally, the remaining 10% of craze growth occurs during the last 10% of the band's cyclic life. It was not possible to monitor changes in craze thickness along with the cycle-dependent changes in craze length. We speculate, however, that the craze thickens slowly at first but reaches its

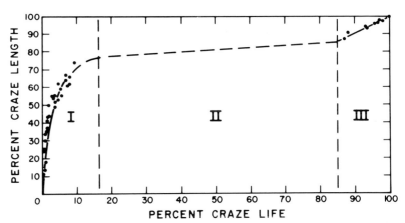

Fig. 4.14 Relationship between band length and band life in PVC-6% DOP under cyclic loading conditions [Reprinted with permission from ASTM STP 675, 1979. Copyright ASTM, 1916 Race St., Philadelphia, PA 19103 (37).]

limiting value (perhaps the critical crack opening displacement at the prevailing stress intensity level) at a more rapid rate when the cyclic life approaches N^*, where N^* is the cyclic life of a given band. As a result, we believe that the somewhat accelerated craze length growth rate during the final 10% of band life is related to an accelerated rate of craze thickening just prior to craze breakdown. To this extent, we suggest that growth of the craze length to the Dugdale dimension l_D represents a necessary but not sufficient condition for fracture. Instead, craze breakdown is believed to be keyed to the rate of craze thickening dt/dN, which is obviously more sluggish than the craze length growth rate dl/dN. The craze aspect ratio l/t, therefore, is thought to increase at first but then decrease with increasing craze length. The experimental findings of Bessonov and Kuvshinskii are consistent with this hypothesis [37, 38].

Gradual thickening of the craze should occur under cyclic loading conditions as the most highly stressed fibrils disentangle and fracture. With the breakage of each additional fibril, the remaining unbroken ligaments are subjected to an ever increasing load and associated strain. These fibrils then become stronger through a process of orientation hardening. The net effect is believed to be the establishment of a relatively constant stress across the craze zone based upon a constant product of fibril strength and remaining fibril volume fraction $\sigma_f v_f$. Thus early in the cyclic life of a discontinuous growth band, a large volume fraction v_f of fibrils with moderate strength σ_f and limited alignment should exist. The failure of a fibril during this early stage of damage zone lifetime should result in minimal load transfer to the remaining fibrils. With continuing fibrillar fracture, the load transferred to the remaining fibrils should increase with these fibrils undergoing further orientation hardening. Finally a critical condition is reached when the next fibril would break; the remaining fibrils would no longer be capable of absorbing the additional load and the entire craze zone would suddenly rupture.

The presumed relative constancy of $\sigma_f v_f$, therefore, reflects a metastable equilibrium condition between fibrillar orientation (strain hardening) and void growth (strain softening), which exists in stage II of the craze length growth process (Fig. 4.14). Stage III would correspond to a condition of approaching instability with the final few fibrils to fracture redistributing an ever-increasing load to the remaining ligaments. By contrast, rapid longitudinal craze development in stage I is believed to reflect an unbalanced condition wherein softening due to craze void formation overwhelms fibril orientation hardening.

It should be noted that the nature of craze thickening kinetics thought to exist in connection with the proposed discontinuous crack growth model, differs from observations made by Verheulpen-Heymans [39, 40] and Kramer and Lauterwasser [41]. These investigators examined the nature of craze thickening in polystyrene and polycarbonate films under monotonic tensile loading conditions and found that the craze aspect ratio and craze strain remained constant along the craze length. From this, it was concluded that craze thickening took place by drawing in new material from the polymer bulk. Though some drawing in of new material from the bulk might be involved in fatigue-generated crazes, we believe that viscoelastic processes such as cyclic-induced craze fibril disentanglement are primarily responsible for craze thickening under cyclic loading conditions. Recalling that the craze grows to 80–90% of its expected length in about 10% of the cyclic life of the band, cyclic-induced fibrillar damage must be taking place within the craze zone so as to cause the craze eventually to break down. This damage, in turn, involves some additional strain to the fibrils and contributes to craze thickening. Were the craze to thicken instead by drawing in fresh material only, there would be no reason for the existing craze fibrils to break down after a certain number of loading cycles. The unresolved question at this time is concerned with the location of this additional craze straining. No evidence has been gathered, thus far, to determine whether the entire craze zone thickens by viscoelastic cyclic-induced fibrillar disentanglement or whether this thickening is confined to the "midrib" region of the craze where strains are found to be greatest [42, 43].

Further studies have shown the discontinuous growth band (DGB) formation process to be a generally observed phenomenon; such markings have been identified in at least seven other polymeric materials (Fig. 4.15) [6, 12, 44]. In all the materials except polyacetal (Delrin and Celcon) and ABS, the

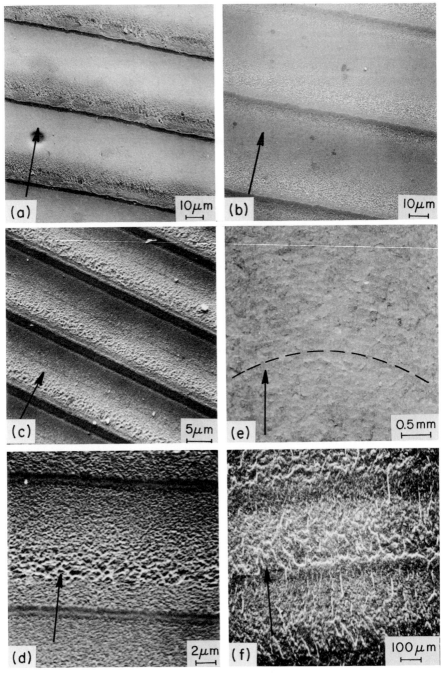

Fig. 4.15 SEM fractographs of discontinuous growth bands in (a) PVC, (b) PS, (c) PSF, (d) PC, (e) PA, (f) ABS. Arrows indicate crack direction [M. D. Skibo, R. W. Hertzberg, J. A. Manson, and S. Kim, *J. Mater. Sci.* **12**, 1977, 531, Chapman and Hall, Ltd.].

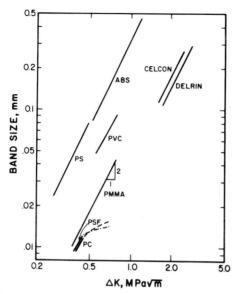

Fig. 4.16 Dependence of band size on ΔK for six amorphous polymers and two grades of polyacetal (Celcon and Delrin) [R. W. Hertzberg, M. D. Skibo, and J. A. Manson, *J. Mater. Sci.* **13**, 1978, 1038, Chapman and Hall, Ltd.].

discontinuous growth bands (DGB) were limited to the mirrorlike regions of their respective fracture surfaces that were generally confined to low ΔK values. At higher ΔK levels, the fracture surfaces were rougher and cloudy in appearance and revealed actual fatigue striations. In fact, some of the striation data shown in Figs. 4.7 and 4.8 were obtained from the very same samples that displayed DG bands. By contrast, the fracture surfaces in polyacetal and ABS were generally cloudier in appearance but did not also reveal individual striations at higher ΔK levels; however, Kitagawa has reported fatigue striations in ABS [24]. Figure 4.16 clearly shows the band width to increase with the second power of the stress intensity factor, consistent with the K dependence of the crack-tip plastic zone size. (The deviation in band width in PC and PSF from a second-power dependence on the stress intensity level is believed to result from an associated transition in the fracture surface appearance from a mirror to a mist texture.) It should be noted that Mills and Walker [45] reported a 1.35 power dependence of DGB spacing in PVC on the stress intensity factor. A close examination of their results, however, reveals that the log scales of their data plot were not identical. When their data are replotted on equivalent log scales, the fracture band spacing does, in fact, vary more closely with the second power of ΔK.

If one assumes that the band width represents the extent of plastic zone development at specific crack length locations, one may infer an apparent yield strength at different K levels by computations based on the Dugdale plastic zone model. Furthermore, if we assume that crack tip yielding occurs by crazing (at least in the amorphous polymers), the inferred yield strength computed from the Dugdale formulation

$$r_y \approx \frac{\pi}{8} \frac{K_{max}^2}{\sigma_{ys}^2} \tag{4.5}$$

would correspond to the tensile stress for craze yielding in the respective polymers [notice that Eq. (4.5) differs little from Eq. (4.1)], where r_y is the crack-tip plastic zone dimension, K_{max} the maximum stress intensity factor, and σ_{ys} the material yield strength. The computation involves setting the plastic zone dimension equal to the band width at a given K_{max} level. The value of K_{max}, in turn, is derived from the ΔK level after correction for the mean stress level. That is, $K_{max} = \Delta K/(1 - R)$, where $R = K_{min}/K_{max}$. It was found in all cases that the computed yield strength was constant at all stress intensity levels (e.g., see Fig. 4.17). Furthermore, the respective yield strengths for each material corresponded to plane strain yield strength or craze stress values previously reported [6].

Further evidence for the existence of the discontinuous cracking process is clearly indicated from the polyacetal test results. Since the spacing between the DGB markings was found to lie in a range of about 0.15–0.35 mm as

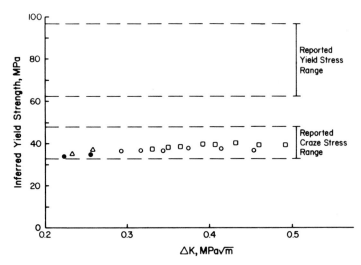

Fig. 4.17 Inferred yield strength computed from discontinuous growth band width and associated ΔK level. Also shown are reported yield and craze strength values for polystyrene [M. D. Skibo, R. W. Hertzberg, and J. A. Manson, *J. Mater. Sci.* **11**, 1976, 479, Chapman and Hall, Ltd.].

compared with the typical increment over which macroscopic crack extension data were collected (i.e., about 0.25 mm) one would expect to find more scatter in the crack velocity test results if the growth bands in PA corresponded to discontinuous crack extension. Indeed, it may be seen from Fig. 4.18 that considerable scatter was associated with the test results for the Delrin and Celcon samples that contained growth bands on the fracture surface while much less scatter was associated with the results from the Delrin sample tested at 10 Hz, where no bands were observed. A similar trend toward increased scatter in FCP test results has been found in ABS at ΔK levels associated with the development of DG band spacings comparable to the usual crack growth interval from one crack-tip reading to the next.

It is instructive at this point to examine the duration of the metastable arrest period associated with the discontinuous crack growth process. The number of loading cycles required for continuous craze development and discontinuous crack extension is estimated rather well by dividing the band width dimension by the corresponding macroscopic crack growth rate. It

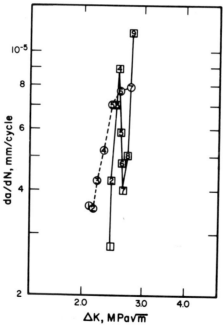

Fig. 4.18 Fatigue crack growth rate data in Delrin with (—□—, 100 Hz) and without (−−○−−, 10 Hz) discontinuous growth bands. Numbers correspond to successive data points. Note scatter in FCP data when bands are present [R. W. Hertzberg, M. D. Skibo, and J. A. Manson, *J. Mater. Sci.* **13**, 1978, 1038, Chapman and Hall, Ltd.].

is clear from Fig. 4.19 that the number of cycles per band decreases strongly with increasing stress intensity level. This most likely reflects the greater extent of specimen damage of each load cycle at higher ΔK levels. Note how the relative ranking of materials in terms of their FCP resistance (Fig. 3.5) parallels the ranking shown in Fig. 4.19 with regard to the DGB cyclic stability. The few exceptions to this relationship (e.g., PVC appearing superior to PC in Fig. 4.19) are related to differences in the material's respective yield strengths. That is, a lower yield strength will generate a larger DGB that would require more loading cycles to break down.

From Fig. 4.19 the cyclic stability N^* is seen to vary with some inverse power of ΔK. Since the band size varies with K^2 and the crack growth rate with K^m, the cyclic life of the DG bands is found to vary inversely with ΔK^{m-2}. Alternatively, $N^* \Delta K^{m-2}$ = constant, which takes on the appearance of the low cycle fatigue relationship for metals proposed by Coffin and Manson [46], i.e., $\Delta \varepsilon_p N^{*c}$ = constant, where $\Delta \varepsilon_p$ is the plastic strain range, and c is a relatively constant value.

It is tempting to extrapolate the experimental results in Fig. 4.16 to determine the ΔK level required to produce a discontinuous growth band in only one loading cycle. In turn, one might ask whether the corresponding band

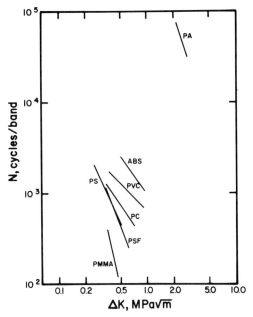

Fig. 4.19 Cyclic stability of discontinuous growth bands in several polymers as function of ΔK [R. W. Hertzberg, M. D. Skibo, and J. A. Manson, *J. Mater. Sci.* **13**, 1978, 1038, Chapman and Hall, Ltd.].

width would be dimensionally consistent with the striation spacing at that extrapolated ΔK level (see Figs. 4.7 and 4.8). In fact, the two fracture band widths are far from equivalent and reflect the fact that two fundamentally different fracture mechanisms are operative at, respectively, different stress intensity conditions. At the extrapolated ΔK levels, the hypothetical DGB widths and actual striation widths differ by two orders of magnitude or more.

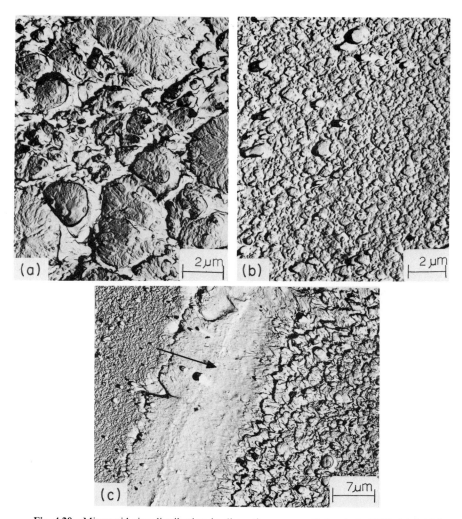

Fig. 4.20 Microvoid size distribution in discontinuous growth bands in PVC. (a) Initial portion of band; (b) final portion of same growth band as in (a); and (c) arrest line separating adjacent bands. Arrow indicates crack propagation direction [R. W. Hertzberg and J. A. Manson, *J. Mater. Sci.* **8**, 1973, 1554, Chapman and Hall, Ltd.].

From the above, it is clear that one must exercise caution when attempting to estimate service conditions based on fracture surface band width measurements, unless the investigator is able to make a clear distinction between DG bands and striations. In addition to the type of macroscopic information discussed earlier, this distinction is made possible with the aid of electron fractographic observations. For the case of PVC, PC, PSF, PS, and PMMA, the discontinuous cracking process occurs within a single craze that developed at the crack tip from the cycling loading conditions [6]. At some point, the crack jumps through the craze before arresting at the craze tip. A close examination of the DGB micromorphology in these amorphous materials reveals the presence of many microvoids that decrease in size in the direction of crack growth. For example, a void diameter gradient from 2 to 0.1 μm is found in DG bands in PVC (Fig. 4.20). This suggests that the cracking process occurs by a void coalescence mechanism with the void size distribution reflecting the internal structure of the craze just prior to crack extension. Note that the observed void size gradient is consistent with the presumed craze opening displacement distribution across the craze. The narrow dark bands (light for the case of PMMA) represent the surfaces of successive crack tips that repeatedly blunted and arrested when the crack reached homogeneous uncrazed material. These narrow zones were delineated by narrow elongated tear dimples that pointed back to the crack origin. Therefore, by direct comparison of DGB and striation micromorphology in homogeneous amorphous polymers (see Figs. 4.6 and 4.15), a clear distinction between these two micromechanisms is possible.

The micromechanisms associated with DGB formation in polyacetal and rubber-modified plastics (ABS and PPO/HIPS) are different from the single-phase amorphous polymers though the phenomenology of the discontinuous crack growth process (i.e., the band size second-power dependence on K) appears to be the same for all engineering plastics. For example, in polyacetal the discontinuous growth bands appear jagged with occasional 50 to 100 μm nodules and cavities dispersed randomly on the fracture surface [12]. These nodules correspond in size to the diameter of individual spherulites. While some secondary cracking has been identified along spherulite boundaries, the advancing fatigue crack generally prefers a transpherulitic path similar to that found in polyethylene at low growth rates [11]. Both types of failure have been reported for brittle failure in a polyacetal (static loading) [47]. For example, the fractograph shown in Fig. 4.21a reflects the passage of the crack through the middle of the spherulite, thereby revealing its radial symmetry. This is compared with a companion SEM micrograph of a polished and etched surface (Fig. 4.21b), which shows the same radial symmetry of a spherulite cut by the polishing plane. Note that a pattern of decreasing microvoid size across each band is observed in neither polyacetal

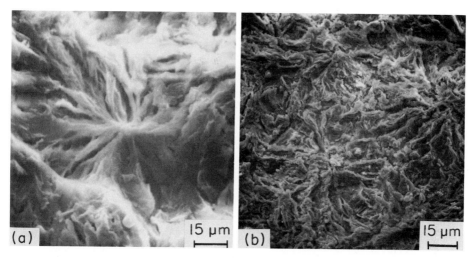

Fig. 4.21 Fatigue fracture in polyacetal (Delrin). (a) Cracking through spherulite; (b) micrograph of polished and etched surface [R. W. Hertzberg, M. D. Skibo, and J. A. Manson, *J. Mater. Sci.* **13**, 1978, 1038, Chapman and Hall, Ltd.].

nor rubber-modified plastics (Fig. 4.15f). For the latter materials, the DGB morphology is very rough and contains numerous patches of material lying on the fracture surface; a slight decrease in surface roughness in the crack growth direction is observed, however.

In an overall sense, the craze fracture model discussed earlier is consistent with several aspects of the DGB formation process. In large measure, the incidence of DG bands depends upon the existence of a single craze (or few crazes at most) at the crack tip. This ensures that cyclic damage will be focused within a narrow zone and would be consistent with the observed fact that DGB formation is limited mainly to low ΔK conditions. (Surely, DGB formation in polyacetal and rubber modified plastics involves some form of multiple crazing. Most likely, this involves a moderate number of crazes that are essentially coplanar rather than multilayered on planes normal to the loading direction.) When ΔK is large, craze bundling develops, cyclic damage is diffused, and DGB formation does not occur [13] (Fig. 4.22). For the same general reason, test frequency would be expected to affect the incidence of DGB formation. Low test frequencies would be expected to allow for additional time for multiple craze nucleation at the crack tip. Since cyclic damage would be diffused, DGB formation at a given ΔK level should decrease with decreasing test frequency [13].

As discussed in Section 3.8.2, the model also anticipates the large polymer FCP sensitivity to M and MWD, previously reported for PS [48], PMMA [49], and PVC [50]. Recall that the craze fracture model envisions craze

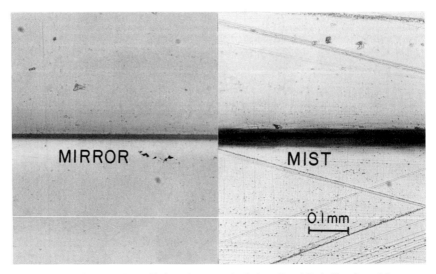

Fig. 4.22   Fatigue crack profile in polystyrene in "mirror" and "mist" regions. Note craze bundling in "mist" region [M. D. Skibo, R. W. Hertzberg, and J. A. Manson, *J. Mater. Sci.* **11**, 1976, 479, Chapman and Hall, Ltd.].

formation to involve fibril formation from only the most highly entangled coils, presumably those containing the longest chains, within the polymer. Consequently, FCP resistance and craze breakdown are seen to be controlled by the long chain M fraction. For example, minor additions of a high-M component to a low-M matrix have resulted in substantial improvement in the FCP resistance of PMMA [51]. This is consistent with previous studies by others [52, 53]. In related fashion, the model accounts for DGB formation only when M is below some critical level (roughly $2–4 \times 10^5$) [6]. When M is high, fibrillar disentanglement proceeds at a very sluggish rate and the polymer tends to form craze bundles that dissipate cyclic damage rather than focus it within a single craze. Consequently, the fatigue crack proceeds through the craze bundle in a continuous manner. Even when M in PMMA is less than about 2×10^5, DGB formation may not occur. If the polymer possesses a bimodal ·MWD with some extra long chain molecules added to the matrix, resistance to cyclic-induced disentanglement may be great enough to suppress the discontinuous crack growth process. For example, Fig. 4.23 shows DGB formation in low-M PMMA but no DGBs in a PMMA sample containing a lower M matrix but with 2 wt. % of a high-M component. It is important to emphasize the fact that the high-M addition changes the average M only slightly.

Additional fractographic observations from samples containing a bimodal MWD provide a reasonable explanation for previous reports of "serrated striations" in low-M PMMA [17, 29]. Janiszewski [54, 55] found that in low-M PMMA samples that contained a small amount of intermediate

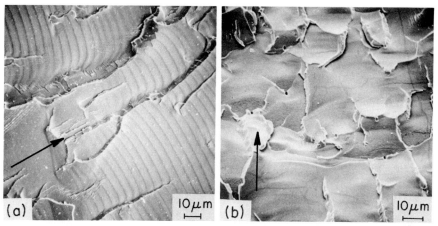

Fig. 4.23 Fatigue fracture surface in PMMA. (a) Low-M PMMA with 0.5% high-M addition. $\Delta K = 0.32$ MPa·m$^{1/2}$. Clearly defined discontinuous growth bands observed; (b) Low-M PMMA with 2% high-M addition. $\Delta K = 0.35$ MPa·m$^{1/2}$. Note absence of discontinuous growth bands. Arrows indicate crack growth direction [Janiszewski, Hertzberg, and Manson (55)].

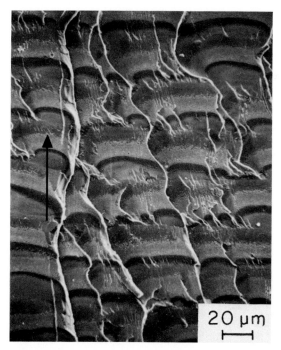

Fig. 4.24 Scalloped discontinuous growth bands in low-M PMMA with 0.5% high-M addition. $\Delta K = 0.41$ MPa·m$^{1/2}$. Arrow indicates crack growth direction [Janiszewski, Hertzberg and Manson (55)].

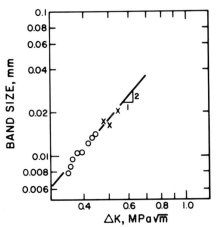

Fig. 4.25   Relationship between discontinuous growth band (○) and scalloped band (×) width and ΔK in low-M PMMA with 1% medium-M addition (LIM) [Janiszewski, Hertzberg and Manson (55)].

M additive, the classical DGB morphology transformed to scalloped DGB markings at higher ΔK levels (Fig. 4.24); the second-power dependence of band size on the stress intensity factor persisted, however (Fig. 4.25). On the basis of these findings, it is reasonable to conclude that the previously reported "serrated striations" were, in fact, representative of a DGB variant known to develop in low-M PMMA.

Recent studies have shown that the morphology of DG bands in PVC also changes with molecular weight; Rimnac *et al.* [56] reported a gradual breakdown in DGB appearance when M_w increased from 97,000 to 210,000 (Fig. 4.26). [Compare these bands with those shown in Fig. 4.15a for a different grade of PVC ($M_w = 108,000$).] Note how the microvoid size gradient within the bands becomes unclear at high M and that the stretch zone is not readily apparent. In marked contrast, these features become much clearer again when a moderate amount of plasticizing agent is added (Fig. 4.26d). Of further interest, no DG bands were found when the M_w was reduced to 67,000.

To summarize, the existence and micromorphology of discontinuous growth bands in engineering polymers appears to depend strongly on M, MWD, and the addition of plasticizing agents.

4.2.3   Additional Observations and Comparisons

To this point, the discussion of polymer fatigue fractographic features has been restricted to the two major sets of markings, fatigue striations

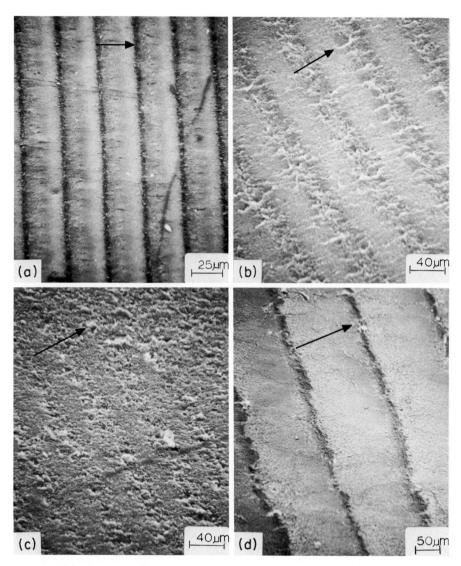

Fig. 4.26 Effect of molecular weight and plasticizer content on discontinuous growth band morphology in PVC. Arrows indicate growth direction. (a) $M_w = 95,000$, $\Delta K = 0.64$ MPa · m$^{1/2}$, (b) $M_w = 170,000$, $\Delta K = 1.0$ MPa·m$^{1/2}$, (c) $M_w = 210,000$, $\Delta K = 1.0$ MPa · m$^{1/2}$, (d) $M_w = 225,000$, $\Delta K = 1.0$ MPa · m$^{1/2}$, 6% DOP plasticizing agent [Rimnac, Hertzberg and Manson (56)].

and discontinuous growth bands, which have been morphologically and quantitatively described. And yet, other fractographic features have been documented but not clearly understood [13]. For example, one finds in PS at higher ΔK levels, though still in the mirror region, that the discontinuous growth bands change abruptly to another series of parallel bands shown in Fig. 4.27. These markings are also found to be oriented perpendicular to the direction of growth. They are similar to the discontinuous growth bands in that they are formed over many load cycles (140–35 per band); however, unlike the discontinuous growth bands, these bands have a relatively constant size of 6–7 μm over a range of growth rates from 5×10^{-5} to 2×10^{-4} mm/cycle. (Similar bands have been found on the fracture surface of PVC samples possessing a relatively low molecular weight, 67,000 [56]. At higher magnification these bands are seen to be composed primarily of small voids whose small variation in size, 0.6–0.2 μm, is responsible for their periodic lineage. The mechanism for the formation of these bands is uncertain at this time and the analysis further complicated by their presence beyond the end of the mirror region and beneath the rough but transparent surface of the mist region. This indicates that these bands represent more than just a surface feature. At still higher ΔK levels within the mist region, the constant width bands disappear with true fatigue striations eventually being formed. The striation mechanism then persists to the point of specimen fracture. Thus, for the case of PS and also PC, a total of three types of parallel fatigue lines are found on the fracture surface, each formed by a different mechanism. (In PVC, two sets of fatigue bands have been identified.) Consequently,

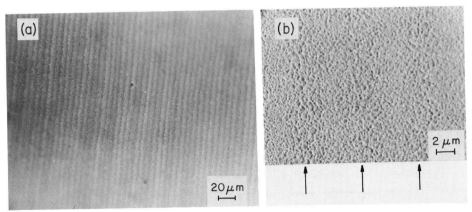

Fig. 4.27   (a) Constant size parallel bands in latter stage of mirror region in polystyrene; (b) note moderate change in microvoid size. Band width given by distance between arrows [M. D. Skibo, R. W. Hertzberg, and J. A. Manson, *J. Mater. Sci.* **11,** 1976, 479, Chapman and Hall, Ltd.].

without a careful electron fractographic investigation or companion macro-scopic fatigue crack growth rate data, it is very difficult to relate safely "growth band spacings" measured on these fatigue fracture surfaces to the advance of the crack during a given load cycle (i.e., to assume that the growth bands were true striations). As such, one cannot automatically separate crack initiation from crack propagation based upon a simple frac-ture band count. That is to say, if an unnotched sample experienced a fatigue life of N cycles with L fatigue lines being found on the fracture surface, the number of loading cycles for crack initiation would not necessarily be $N - L$. The latter would be true if the fatigue lines were, indeed, striations. However, if the fatigue lines were DG bands or the constant width lines, the extent of the crack propagation stage would be hundreds or even thousands of times greater than that based on a simple fatigue line count.

Computations of this type had been reported previously by Havlicek and Zilvar [31]. Initially, PS fatigue lines were interpreted as being striations that suggested a crack initiation stage consuming over 99% of the sample life. In a subsequent report, Zilvar *et al.* [25] correctly reinterpreted these fatigue lines as being those associated with the discontinuous crack growth process; accordingly, the relative extent of the initiation and propagation stages significantly changed. Documentation of DGB formation in PS un-notched fatigue samples has also been reported by Sauer *et al.* [35].

It is appropriate, therefore, to reexamine the conclusions drawn by Rabinowitz *et al.* [57] regarding their fractographic studies of fatigued-unnotched PS samples. They observed in region III of the fatigue process in PS (see Fig. 2.19c) that the initial part of the fracture surface was relatively featureless with the exception of a series of crack arrest lines, which they interpreted as being individual fatigue striations (region ABCA in Fig. 4.28). The spacing between these lines was found to be on the order of 8 μm. Assuming these lines to be striations, they would represent too high a crack growth rate since the specimens in this test regime (i.e., region III) had experienced fatigue lives between 10^5 and 10^7 cycles. Instead, if one assumes an appropriate cyclic stress range [derived from the σ-N curve for the material (see Fig. 2.19c)] and a reasonable estimate of the crack length associated with the fatigue band photograph shown in their paper, a K level of approx-imately 0.2–0.3 MPa·m$^{1/2}$ is computed in association with the fatigue lives. It is important to note that this value is comparable to the K level where DGB lines were recorded in our fatigue crack propagation experiments. Combining this with the additional confirmation of DGB formation in PS by Havlicek *et al.* [25] and Sauer *et al.* [35], we conclude that the lines found in region III by Rabinowitz *et al.* [57] were, in fact, discontinuous growth bands which are uniquely dependent upon prior craze development. [Most probably, the fracture bands in PC shown by McEvily *et al.* [20] (see their

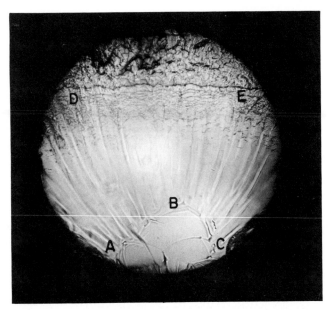

Fig. 4.28   Fatigue fracture appearance in region III of unnotched polystyrene. Fatigue crack covers area ADECA [S. Rabinowitz, A. R. Krause and P. Beardmore, *J. Mater. Sci.* **8**, 1973, 11, Chapman and Hall, Ltd.].

Fig. 2) also are DG bands and not fatigue striations as originally stated.] In related fashion, the sudden transition in the fracture surface appearance that Rabinowitz *et al.* observed probably represented the point beyond which multiple crazing occurred and the DGB process terminated [13]. In this regard, a rough calculation of the stress intensity level associated with the crack front fracture surface transition point reveals K to be in the range of approximately 0.3–0.4 MPa·m$^{1/2}$, which corresponds to the transition point where the DGB process terminated [13]. Finally, Beardmore and Rabinowitz reported the presence of additional lines on the fracture surface with a spacing on the order of 10 μm. The fact that they did not observe an increase in the spacing of these lines with increasing crack length leads us to believe that these lines may very well have been the constant spacing lines that form at intermediate K levels.

This discussion is certainly not intended to detract from the extensive and illuminating fatigue results reported by previous investigators, especially Rabinowitz *et al.* [57] (see Chapter 2), but rather to focus attention on the problem of accurate interpretation of polymer fatigue fracture surface markings.

The fracture surface micromorphology of rubber-modified plastics invites final comment. As discussed in Section 5.1.2, rubber additions to glassy

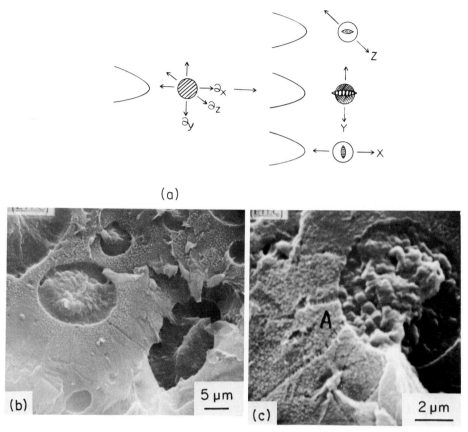

(a)

(b)

(c)

5 µm

2 µm

Fig. 4.29 Craze development in high impact polystyrene. (a) Model showing possible locations for craze development due to triaxial stress field; (b) fatigue fracture surface ($da/dN = 2.5$ µm/cycle) revealing ruptured rubber particles and equatorial crazes evidenced by void formation (white halo) at perimeter of rubber particle; (c) secondary craze at A growing from rubber particle revealing oriented fibrils normal to plane of craze ($da/dN = 0.25$ µm/cycle) [J. A. Manson and R. W. Hertzberg, *J. Polym. Sci.*, Polymer Physics Ed. **11**, 1973, 2483. Reprinted by permission of John Wiley & Sons, Inc.].

polymer matrices provide for more extensive localized crazing; this results in improved fracture toughness and fatigue crack propagation resistance. In terms of the local three-dimensional stress field surrounding each rubber particle, it is possible to develop a complex network of crazes oriented nominally perpendicular to each principal stress axis (Fig. 4.29a). Evidence for crack growth through preexistent equatorial crazes surrounding a rubber particle in high-impact polystyrene may be seen in Fig. 4.29b. The light halos surrounding the rubber particles represent clusters of microvoids formed during the initial craze development stage [58]. Evidence for fatigue-induced craze formation oriented parallel to the applied stress direction is seen in Fig. 4.29c. Note clear evidence of fibril formation normal to the plane of the craze (see region A).

REFERENCES

[1] A. Philips, V. Kerlins, and B. V. Whiteson, Electron Fractography Handbook, AFML TDR-64-416, WPAFB, Ohio (1965).
[2] "Metals Handbook," Vol. 9. American Society of Metals, Metals Park, Ohio, 1974.
[3] T. P. Rich and D. J. Cartwright (eds.), "Case Studies in Fracture Mechanics," AMMRC MS 77-5, Watertown, Massachusetts, 1977.
[4] R. W. Hertzberg, "Deformation and Fracture Mechanics of Engineering Materials." Wiley, New York, 1976.
[5] J. A. Manson and R. W. Hertzberg, *CRC Crit. Rev. Macromol. Sci.* **1**(4), 433 (1973).
[6] M. D. Skibo, R. W. Hertzberg, J. A. Manson, and S. Kim, *J. Mat. Sci.* **12**, 531 (1977).
[7] J. P. Berry, *Nature (London)* **185**, 91 (1960).
[8] R. P. Kambour, *J. Polym. Sci. Part A* **3**, 1713 (1965).
[9] R. P. Kambour, *J. Polym. Sci. Part A-2* **4**, 349 (1966).
[10] P. E. Bretz, Ph.D. Dissertation, Lehigh University (1980).
[11] E. H. Andrews and H. J. Walker, *Proc. R. Soc. London Ser. A* **325**, 57 (1971).
[12] R. W. Hertzberg, M. D. Skibo, and J. A. Manson, *J. Mat. Sci.* **13**, 1038 (1978).
[13] M. D. Skibo, R. W. Hertzberg, and J. A. Manson, *J. Mat. Sci.* **11**, 479 (1976).
[14] R. C. Bates and W. G. Clark, Jr., *Trans. Q. ASM* **62**(2), 380 (1969).
[15] F. A. McClintock, ASTM STP 415, p. 170 (1967)
[16] E. H. Andrews, *J. Appl. Phys.* **32**(3), 542 (1961).
[17] C. E. Feltner, *J. Appl. Phys.* **38**, 3576 (1967).
[18] N. E. Waters, *J. Mat. Sci.* **1**, 354 (1966).
[19] N. H. Watts and D. J. Burns, *Polym. Eng. Sci.* **7**, 90 (1967).
[20] A. J. McEvily, R. C. Boettner, and T. Johnston, "Fatigue, An Interdisciplinary Approach" (J. J. Burke, N. L. Reed, and V. Weiss, eds.). Syracuse Univ. Press, Syracuse, New York, 1964.
[21] G. H. Jacoby, "Electron Microfractography," ASTM STP 453, p. 147 (1969).
[22] L. J. Broutman and S. K. Gaggar, *Int. J. Polym. Mat.* **1**, 295 (1972).
[23] G. Jacoby and C. Cramer *Rheol. Acta* **7**, 23 (1968).
[24] A. Misawa, M. Takasi, and T. Kunio, *16th Japan Cong. on Mat. Res.*, **16**, 207 (1973).
[25] V. Zilvar, V. Havlíček, and P. Bouška, *Proc. Conf. Dimension. Strengthen. Calculations, Budapest* II-471 (1974).
[26] M. Kitagawa, *Bull. J. Soc. Mech. Eng.* **18**(117), 240 (1975).
[27] V. I. Singian, J. W. Teh, and J. R. White, *J. Mat. Sci.* **11**, 703 (1976).
[27a] J. R. White and J. W. Teh, *Polymer* **20**, 764 (1979.
[28] B. Tomkins and W. D. Biggs, *J. Mat. Sci.* **4**, 532 (1969).
[29] T. A. Johnson, *J. Appl. Phys.* **43**(3), 1311 (1972).
[30] C. Laird and C. G. Smith, *Phil. Mag.* **7**, 847 (1962).
[31] V. Havlíček and V. Zilvar, *J. Macromol. Sci.* **B5**, 317 (1971).
[32] J. P. Elinck, J. C. Bauwens and G. Homès, *Int. J. Fract. Mech.* **7**(3), 227 (1971).
[33] R. W. Hertzberg and J. A. Manson, *J. Mat. Sci.* **8**, 1554 (1973).
[34] T. Kurobe and H. Wakashima, *J. Soc. Mat. Sci.* **21**(227), 800 (1972).

[35] J. A. Sauer, A. D. McMaster, and D. R. Morrow, *J. Macromol. Sci.-Phys.* **B12**(4), p. 535 (1976).

[36] D. S. Dugdale, *J. Mech. Phys. Solids* **8**, 100 (1960).

[37] R. W. Hertzberg, M. D. Skibo and J. A. Manson, ASTM STP 675, p. 471 (1979).

[38] M. I. Bessonov and E. V. Kuvshinskii, *Sov. Phys.-Solid State* **3**(5), 950 (1961).

[39] N. Verheulpen-Heymans, "Deformation, Yield and Fracture of Polymers," p. 35-1. Plastics and Rubber Institute, London, 1979.

[40] N. Verheulpen-Heymans, *Polymer* **20**, 356 (1979).

[41] E. J. Kramer and B. D. Lauterwasser, "Deformation Yield and Fracture of Polymers," p. 34-1. Plastics and Rubber Institute, London, 1979.

[42] P. Beahan, M. Bevis, and D. Hull, *Polymer* **14**, 96 (1973).

[43] P. Beahan, M. Bevis, and D. Hull, *Proc. R. Soc. London Ser. A* **343**, 525 (1975).

[44] M. D. Skibo, J. Janiszewski, R. W. Hertzberg, and J. A. Manson, "Toughening of Plastics," p. 24.1. Plastics and Rubber Institute, London, 1978.

[45] N. J. Mills and N. Walker, *Polymer* **17**(4), 335 (1976).

[46] L. F. Coffin, *Trans. ASME* **76**, 931 (1954).

[47] C. F. Hammer, T. A. Koch, and J. F. Whitney, *J. Appl. Polymer. Sci.* **1**, 169 (1959).

[48] J. A. Sauer, E. Foden, and D. R. Morrow, *SPE Tech. Papers* **22**, 107 (1976).

[49] S. L. Kim, M. D. Skibo, J. A. Manson, and R. W. Hertzberg, *Polym. Eng. Sci.* **17**(3) (1977).

[50] M. D. Skibo, R. W. Hertzberg, J. A. Manson, and E. A. Collins, *Proc. Int. Conf. PVC, 2nd, Lyon* p. 233 (1976).

[51] S. L. Kim, J. Janiszewski, M. D. Skibo, J. A. Manson, and R. W. Hertzberg, *ACS Organic Coatings Plast. Chem.*, **38**(1), 317 (1978).

[52] S. Wellinghoff and E. Baer, *J. Macromol. Sci.* **B11**, 367 (1976).

[53] J. F. Fellers and B. F. Kee, *J. Appl. Polym. Sci.* **18**, 2355 (1974).

[54] J. Janiszewski, M. S. Thesis, Lehigh Univ. (1978).

[55] J. Janiszewski, R. W. Hertzberg and J. A. Manson, Manuscript in preparation.

[56] C. M. Rimnac, R. W. Hertzberg, and J. A. Manson, ASTM STP XXX, (1980).

[57] S. Rabinowitz, A. R. Krause, and P. Beardmore, *J. Mat. Sci.* **8**, 11 (1973).

[58] J. A. Manson and R. W. Hertzberg, *J. Polym. Sci. Polym. Phys. Ed.* **11**, 2483 (1973).

[59] G. Pitman and I. M. Ward, *J. Mater. Sci.* **15**, 635 (1980).

Note added in Proof: Pitman and Ward [59] recorded shear lip width measurements in polycarbonate as a function of specimen thickness and ΔK level.

5 | *Composite Systems*

Since ancient times, man has combined materials together in the form of laminates, dispersed particles or fibers into a matrix material [1–4]. By doing so, it was possible to obtain combinations of strength or toughness with other properties that could not be found with any single component alone. Thus, cementitious concretes, laminated archery bows, bricks, mortars reinforced with straw or hair, and dispersion-hardened metals have been known for centuries; many natural products that have an excellent balance of properties are also composites, for example, wood and bone.

As modern technology has developed, so has the interest in new man-made composites, many of which include a polymer as one of the components. Practical interest has been stimulated not only by the desire for improved strength, stiffness, or toughness, but also by the fact that in many cases excellent properties can be obtained in materials whose densities are much lower than those of traditional materials such as metals. The high "strength-to-weight" or high "stiffness-to-weight" ratios that have been achieved have become increasingly important in aerospace and other structural applications in which the saving of weight is greatly to be desired. More recently, the use of composites in engineering materials for less critical applications has increased because their use may imply lower consumption of fuels in vehicles, lower energy costs per unit of performance relative to many other materials, and conservation of nonrenewable raw materials. At the same time, many composites have high levels of mechanical properties, including in some cases exceptional resistance to certain kinds of fatigue loading.

Apart from technological importance, composites are also of fundamental concern. While their properties are frequently defined by some average of constituent properties, interactive effects are frequently present. Thus, the characterization of composite behavior is complicated in interesting ways by the fact that even a nominally two-component system contains interfaces and even discrete interfacial regions whose properties are often difficult to infer [3, Chap. 12]. Additional complications, some peculiar to fatigue loading, arise due to the viscoelastic (and hence time-dependent) nature of polymers. Hence, fundamental understanding of the fatigue behavior of polymeric composites poses a strong challenge to the scientist.

This chapter is concerned with several principal types of polymer composites: (1) plastics containing a dispersed toughening phase (usually rubbery); (2) plastics containing particulate fillers; and (3) plastics containing discontinuous or continuous fibrous reinforcements. Discussion is limited to man-made composites, and, as indicated, to more or less rigid matrixes (for reviews of fatigue in reinforced elastomers, see Beatty [5] and Manson and Hertzberg [6]). However, because of their present or potential interest as structural materials, fatigue in several systems not usually thought of as composites, adhesive joints and polymer-concrete materials, is also discussed, at least briefly.

5.1. TOUGHENED PLASTICS

5.1.1. Background

As polystyrene began to be introduced into commercial applications during the 1940s, it became apparent that its low impact strength posed an increasingly serious limitation on use. Indeed, as shown in Table 5.1, low impact strengths are characteristic of many unmodified polymers, not only those that are glassy, amorphous, and brittle at room temperature, but also those that are semicrystalline and, though normally more or less ductile, notch-sensitive.*

One intuitive and historical approach to the toughening of a brittle polymer is to make use of the principle of composite action, that is, to combine a brittle but otherwise desirable material with a separate, inherently tough component. Such an approach was taken over 50 yr ago by Ostromi-

---

* Thus, even with a polymer such as polypropylene that is normally ductile when unnotched, the triaxiality of the stress field associated with a notch tip in a reasonably thick specimen often results in raising the yield stress with respect to the stress for brittle fracture. This increase is accentuated by the high strain rates associated with impact loading so that fracture occurs prior to yielding [7, p. 299].

TABLE 5.1   Typical Impact Strengths[a] of Plain and Toughened Plastics

Polymer system	Impact strength (ft lb/in.)[b] at room temperature
Polystyrene (PS)	0.2–0.4
HIPS[c]	0.5–8
ABS	1–12
Poly(methyl methacrylate) (PMMA)	0.3–0.5
HI–PMMA	0.8–2.5
Poly(vinyl chloride) (PVC)	0.4–1
HI–PVC	10–15
Phenolic resin	0.2–0.4
Epoxy resin (E)	0.2–1
HI–epoxy	~10
Polycarbonate (PC)	12–18
PC–ABS alloy	8–13
Polypropylene (PP)	0.5–2
HI–PP	1–20
Poly(phenylene oxide) (PPO)	1.5–1.8
HIPS-PPO alloy	5–8

[a] By the notched Izod test, with data ranges taken from the Modern Plastics Encyclopedia [9]. In some cases, nonstandard dimensions may be involved.

[b] Conversion: 1 ft lb/in. = 0.53 J/cm.

[c] The prefix "HI" is used here and subsequently to indicate materials modified to improve impact strength.

slensky [8], who patented a process for polymerizing a solution of rubber in styrene monomer. Although the products of this process had other inherent limitations, the concept was valid, and subsequent developments led to the commercialization of "high-impact" polystyrene (HIPS) beginning in the late 1940s. While the direct mechanical blending of rubbery materials such as styrene–butadiene copolymers was also introduced, solution polymerization techniques continued to offer advantages, and, by the 1960s, came to dominate production. For example, a degree of grafting is developed in the solution-polymerized materials between the rubber and the polystyrene, thus improving the bonding of the rubbery inclusions to the matrix; in addition, the effective volume of the rubbery particle is increased, thus conferring improved toughness.

Since then many improvements have been made in HIPS, and, even more importantly, the basic concept of introducing a toughening phase (whether a classical rubber or not) has been applied to many other polymers. Thus (see Table 5.1) not only have other toughened polystyrenes such as acrylonitrile–butadiene–styrene resins (ABS) appeared on the market, but also new multiphase polymers have been developed based on, for example, poly(vinyl chloride) (PVC), poly(methyl methacrylate) (PMMA), polypropylene (PP) nylon 66, poly(2,6-dimethyl-1,4-phenylene oxide) (PPO), polyacrylonitrile, polycarbonate (PC), and epoxy resins. In some cases, e.g., PVC, a classically rubbery phase is introduced; in others, the base polymer is "alloyed" with another tough one, sometimes to reduce cost or improve processability. Thus PPO is alloyed with HIPS, and polycarbonate with ABS resin. All these products are finding significant markets, and toughened epoxies are playing an increasingly important role in structural adhesives. As time passes, further similar improvements in the toughness of other brittle or notch–sensitive polymers may be expected. While, as expected from simple composite theory, the stiffness and ultimate strength of a resin is decreased by the incorporation of a rubbery phase, some sacrifice is worthwhile to gain toughness. With nonrubbery inclusions, a major sacrifice may not be necessary.

Since a general relationship between fracture toughness and fatigue resistance has been demonstrated, it is useful to consider the general question of toughening (see Section 3.8.7). Thus, in the following sections, the principle features of toughening by the inclusion of second phases are outlined, and implications with respect to fatigue behavior are presented and discussed. For discussion of plastics containing inorganic or high-modulus organic particulate fillers, see Section 5.2.

5.1.2. General Mechanisms of Toughening

Although many questions about toughening in rubber-modified plastics remain to be answered in detail, considerable insight has been gained within the past decade. (For detailed reviews see Kambour [10], Mann and Williamson [11], Keskkula *et al.* [12], Manson and Sperling [3, Chap. 3] and especially the monograph by Bucknall [7].) It now seems clear that the stress concentrations developed in tension (Figs. 4.28 and 5.1) at the particle–matrix interfaces result in a combination of crazing and shear yielding in the matrix. The net result is that damage is highly delocalized, and much energy is therefore dissipated over a large volume. In addition, the growth of small crazes [13] is inhibited by interactions with each other or with shear bands (Figs. 5.2 and 5.3), and rubber particles themselves may physically serve as craze stoppers. Another possibility is that the presence of a low-modulus rubbery phase may, by decreasing the capability of stress transfer in the z direction (the through-thickness direction) (Fig. 4.28), alleviate the severity of the stress state in the matrix and encourage the relative role of energy dissipation in shear. In any case, the principal contribution to fracture toughness arises from the response

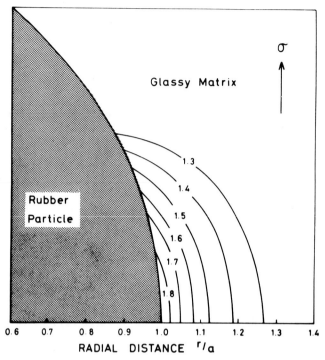

Fig. 5.1 Contour map showing stress concentrations, calculated using Goodier's equations, near the equator of an isolated rubber particle in HIPS under uniaxial tension [Bucknall (7, p. 129)].

of the *matrix*; the deformation of the rubber itself contributes only a small fraction of the total.

Regardless of the precise balance of mechanisms that may hold for a particular case, the effect on toughness depends on the concentrations of the rubber and the dispersed phase (which normally contains both rubbery and glassy constituents), the particle size, and the interfacial bonding. The properties of the rubbery phase *per se* are also important, for maximum energy dissipation can occur only when the nominally rubbery phase is in fact rubbery under the test conditions concerned. For example, at the high strain rates characteristic of impact loading, glass transition temperatures (T_gs) may be increased by as much as 60°C, so that a relatively high-T_g elastomer may well become glassy, and hence ineffective as a toughening agent, on impact.

According to Bucknall [7, Chap. 8], crazing is likely to be the dominant response to tensile loading the higher the strain rate and the larger the particle size of the rubbery phase. Also, with notched specimens, the triaxiality at the notch tip appears to be reduced, so that with notch-sensitive but otherwise ductile materials (e.g., PVC or PP), advantage can be taken of their

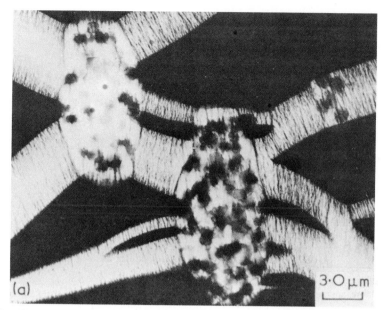

Fig. 5.2 Crazes connecting rubber particles in a cast thin film of HIPS, which was strained in tension, stained with osmium tetroxide, and examined in the transmission electron microscope [P. Beahan, A. Thomas, and M. Bevis, *J. Mater. Sci.* **11**, 1297 (1976), Chapman & Hall Ltd.].

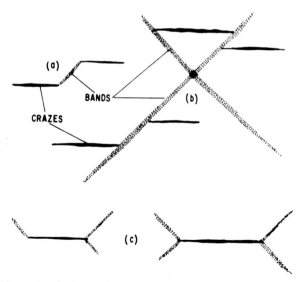

Fig. 5.3 Schema showing interactions of crazes with each other and with shear bands, based on observations of micrographs of several authors cited by Kambour. [After R. P. Kambour, J. Polym. Sci., *Macromol. Rev.* **7**, 1 (1973). Copyright © 1973, by John Wiley & Sons, Inc. Reprinted by permission of John Wiley & Sons, Ltd.].

inherent ductility. In other words, with ductile materials the yield envelope is not far above the crazing envelope [7, Chap. 6], even on impact loading, so that crazing is accompanied by a significant degree of shear yielding.

5.1.3 Implications with Respect to Fatigue

There are several implications of rubber toughening with respect to fatigue:

(1)   The balance between shear yielding and crazing is known to depend on the strain rate; the higher the strain rate, the greater the relative contribution of crazing.

(2)   The modulus is decreased, and the creep rate is increased, in proportion to the volume fraction of the rubbery phase.

(3)   The rubbery phase increases the loss modulus in proportion to its volume fraction, and, at the same time, crazing increases the loss modulus even more.

The viscoelastic and plastic response may be expected to result in complex effects of cyclic wave form and frequency, which will affect the strain rate, time-under-load, and hysteretic heating. As will be seen below, this expectation is confirmed.

Shear Yielding versus Crazing.   In a series of elegant experiments, Bucknall *et al.* [7, pp. 14–16] have clearly demonstrated how the balance between yielding and crazing in a rubber-modified polymer is reflected in

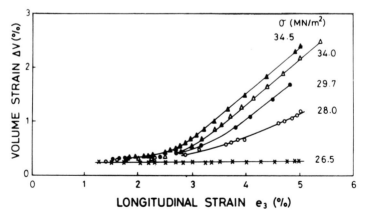

Fig. 5.4   Relationship between volumetric and linear creep (ΔV and e_3, respectively) for ABS at 20°C showing change in the dominant mechanism of tensile creep from shear to crazing with increased stress and strain [C. B. Bucknall and J. C. Drinkwater, *J. Mater. Sci.* **8**, 1800 (1973), Chapman & Hall Ltd.].

creep. By measuring *volumetric* creep (which will occur only as a result of dilation, and hence crazing) simultaneously with *longitudinal* creep (which is, of course, a consequence of shear deformation), it is possible to measure the relative contribution of crazing versus shear yielding (see Figs. 5.4 and 5.5). Thus the slope of the creep curve gives a direct measure of the balance: slopes of unity and zero correspond to 100% crazing and 100% shear yielding, respectively. Application of this technique shows, for example, that HIPS deforms essentially by crazing at *all* strains (Fig. 5.5). In contrast, a typical ABS (Fig. 5.4) deforms predominantly by shear yielding at low stresses and strains, but with an increasing component of crazing as stresses and strains are increased (up to 85% crazing in the upper curve of Fig. 5.4). While toughened PVC also tends to deform at low stresses and strains mainly by shear yielding, quantitative extension of creep tests to very high strain rates has not been achieved. Nevertheless, it is likely that the balance should be shifted toward crazing under such loading. Interestingly, even on impact, ABS exhibits a higher fracture energy than HIPS (Table 5.1) an observation suggesting that even low-strain-rate tests can help rank the efficiency of energy-dissipation processes.

In any case, the role of creep in fatigue loading (and, in particular, the

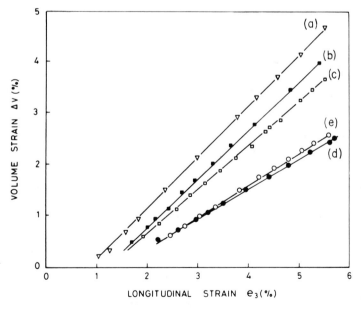

Fig. 5.5 Effects of matrix composition on creep mechanism in HIPS/PS/PPO blends at 20°C. Compositions are as follows: (a) 50/50/0; (b) 50/37.5/12.5; (c) 50/25/25; (d) 50/12.5/37.5; (e) 50/0/50 [Bucknall (7, p. 220)].

balance between shear and crazing response) must depend at least on the waveform and frequency, and probably on mean stress as well. Since shear response involves greater inherent energy dissipation than crazing, the indirect effect on fracture toughness may well be significant.

Modulus and Creep.   As expected (Fig. 5.6), the modulus of a rubbery phase–glassy matrix composite is decreased in proportion to the volume fraction of rubbery phase [17]. Clearly the effect of this is to lower the constraint imposed by the matrix on the growth of a crack. At the same time, the rates of creep due to both crazing and shear stress components are significantly increased; this observation has been interpreted in terms of an increase on the stress concentration factor in the Eyring creep equation, which governs creep in these as well as many other polymer systems [7, Chap. 8].

Damping and Hysteretic Heating.   As shown in Fig. 5.6, the expected decrease in the shear modulus of polystyrene with increasing rubber content is accompanied by significant changes in the damping spectrum.* Thus a peak in the loss tangent appears at $\sim -90°C$, reflecting the T_g of the added polybutadiene; as expected, the magnitude of the peak increases with increasing rubber content. Even more important from the standpoint of fatigue, the loss tangent increases with the rubber content. The $\sim 25\%$ increase in tan δ at room temperature caused by the presence of 10% rubber (corresponding to about a twofold increase in loss modulus) could be quite significant in the induction of hysteretic heating during cycling between fixed load limits (see Section 2.3.2); since the tan δ curve is rising at room temperature, hysteretic heating will develop in an autoaccelerating manner.

Rubber content *per se* is not the only factor affecting the damping spectrum. The actual rubbery phase in a solution-grafted HIPS also contains subinclusions of polystyrene (Fig. 5.2); in fact, the volume fraction of

---

* While dynamic mechanical spectroscopy directly gives information about the storage and loss of energy (damping) during oscillation of a specimen, similar information may, in principle, be derived from static experiments. Thus, J' and J'' may be derived from the relaxation spectrum of a sample, since static and dynamic responses are related by Fourier transforms (17a, p. 73):

$$G'(\omega) = G_e + \omega \int_0^\infty [G(t) - G_e] \sin \omega t \, dt,$$

$$G''(\omega) = \omega \int_0^\infty [G(t) - G_e] \cos \omega t \, dt,$$

where $G'(\omega)$ and $G''(\omega)$ are the loss and storage shear moduli as functions of the frequency ω (in radians/sec), $G(t)$ is the relaxation modulus as a function of time t, and G_e is the equilibrium modulus. Analogous equations hold for the complex compliance as related to static creep compliance (see Ref. 17a).

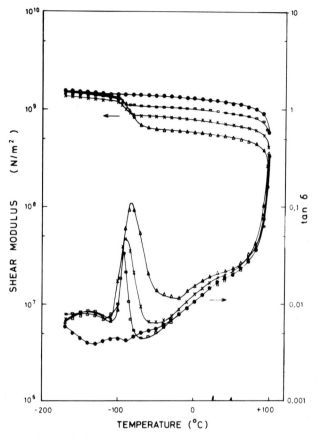

Fig. 5.6 Dynamic mechanical loss curves for styrene polymers showing secondary loss peaks in HIPS at $\sim -90°C$ due to glass transition of polybutadiene: (●) polystyrene containing no rubber; (□) HIPS made by blending PS mechanically with 10% rubber; (×) HIPS bulk polymer containing 5% rubber; (△) HIPS bulk polymer containing 10% rubber. [G. Gigna, *J. Appl. Polym. Sci.* **14,** 1781 (1970). Copyright © 1970 by John Wiley & Sons, Inc. Reprinted by permission of John Wiley & Sons, Inc.]

rubbery phase (as distinct from *volume fraction of the rubber alone*) plays a role as well. Such a role is exemplified in Figs. 5.6 and 5.7. In the former it is seen that at constant rubber content (10% in this case), the higher volume fraction of dispersed phase in the solution-grafted resin (due to the inclusion of the PS subphase) yields the higher values of tan δ. In Fig. 5.7, it is seen that, at high volume fractions of the dispersed phase, the damping curves are shifted vertically, the major peaks are shifted to higher temperatures, and new peaks occur. Depending on the temperature in such a system, hysteretic heating could be either autoaccelerating or not. Other factors such as

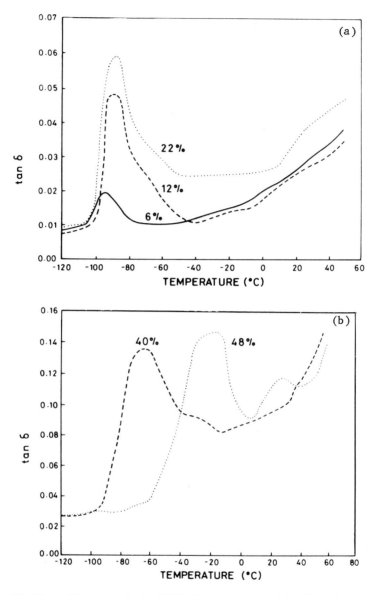

Fig. 5.7 Dynamic loss curves for five HIPS polymers, each containing 6% polybutadiene, but with volume loadings of rubber particles between 6 and 48%, owing to differences in the concentrations of polystyrene subinclusions within the composite rubber particles [E. R. Wagner and L. M. Robeson (18)].

morphology and cross-linking may also serve to perturb the dynamic spectrum [18].

In any case, the occurrence of hysteretic heating may be expected to change the viscoelastic state by softening of whatever volume is subjected to the heating (from the plastic zone to the whole specimen, depending on the volume experiencing maximum stress), with autoacceleration in the case of constant-load-range cycling if E'' is increasing with temperature above ambient. There may be other effects as well, for example, an enhancement in the rate of craze recovery on the unloading part of the strain cycle (see Bucknall [7, p. 240] and Cessna [19]). Also, a rise in the local temperature may well affect the balance between crazing and shear deformation.

5.1.4. Crazing and Shear in Fatigue

Although little detailed information is available, it seems likely that some combination of crazing and shear response exists during fatigue as well as in static tests. As shown in Fig. 4.28, clear evidence exists for the occurrence of crazing during fatigue crack propagation in HIPS. On the fracture surfaces, which are at right angles to the loading direction (i.e., in the x direction), bands of whitened nodules may be seen around the rubber particles; the bands also overlap in some cases. This appearance may be interpreted in terms of equatorial crazing in the matrix about the rubber particles ahead of the crack tip (i.e., in the x direction and xz plane, as expected for crazing), followed by rupture of the craze fibrils as the crack passes through the craze. The occasional observation of crazes in other planes (e.g. the xy plane, Fig. 5.1) is also to be expected for two reasons. First, the stress in the z direction is not zero, and, second, the orientation of the stress field itself may be perturbed by the inhomogeneity of the material. Since no evidence for shear bands is seen, it is likely that crazing is the dominant mechanism underlying fatigue crack propagation in HIPS. Evidence for a major role of crazing in the fatigue of HIPS has also been found by Bucknall and Stevens [20], and by Bartesch and Williams [21]. These observations are consistent with prior predictions of Bucknall [7, Chap. 8] for static loading at both low and high rates.

Of course, if the effects of hysteretic heating dominate fatigue failure, shear response may be expected to predominate. Thus with unnotched ABS containing insignificant surface flaws, Bucknall and Stevens [20] showed fatigue failure to be ductile in nature; in contrast, the presence of significant surface flaws resulted in the "brittle" propagation of a crack.

Evidence for a major role of shear in fatigue crack growth has been observed in rubber-toughened PVC. In Fig. 5.8, it is clear that, as with static

Fig. 5.8 Scanning electron micrograph of fracture surface of PVC ($M_w = 2.1 \times 10^5$) toughened by the addition of a methacrylate-butadiene-styrene rubber (14 phr, parts per hundred of resin.) Note the evidence for a high degree of matrix drawing (not seen in the unmodified PVC) induced by the rubbery phase (which cannot be seen under these conditions of observation). (Micrograph at 1200× in the fast fracture region, with crack direction from bottom to top, by C. Rimnac, Lehigh University.)

loading, the presence of the rubber particles has stimulated a vigorous shear response in the somewhat ductile matrix. While the occurrence of crazing cannot be excluded, the shear response appears to be dominant in this case. Even if crazing has occurred, it is possible that especially rapid recovery of crazes may take place with an inherently ductile polymer; indeed, Cessna [19] has suggested that such rapid recovery may be responsible for the failure to observe crazes in rubber-modified PVC broken under static loading. Thus even when the damage is limited to the plastic zone rather than diffused throughout the volume of the specimen (that is, at low values of ΔK, low frequencies, and low rubber contents), some balance between crazing and shear deformation may be expected.

5.1.5. Fatigue Behavior

Failure in Unnotched Specimens. The importance of crazing in deter-mining yield behavior and mechanical loss was dramatically illustrated in a series of cyclic stress–strain experiments by Bucknall [22]. As shown in Fig. 5.9, a sequence of stress–strain cycles on a single unnotched specimen of HIPS (with stresses high enough to induce yielding and stress-whitening) revealed progressively greater changes in response. As cycling proceeded, the polymer exhibited increased softening and larger hysteresis loops. Also, a second, lower yield stress may be noted, the magnitude decreasing with cycling. Crazing is associated with the upper yield point, and yielding with the lower [10]; in fact, the initial portion of the hysteresis curve resembles a typical curve obtained by Kambour and Kopp [23] for the deformation of a single craze in polycarbonate. It is interesting to note the contributions of various components of work per cycle deduced by analysis of the deformation curves [24]: energy of plastic deformation, 65%; viscous energy, 20%; and recoverable elastic energy, 15%. These proportions support the view that the major contribution to fracture toughness in HIPS is associated with deformation of the matrix. At the same time, considerable recovery takes place on unloading as the crazes close up (craze "healing") so that crazing implies not only plastic deformation but a significant recoverable viscoelastic element as well.

Effects of constant-load cycling were examined by Bartesch and Williams

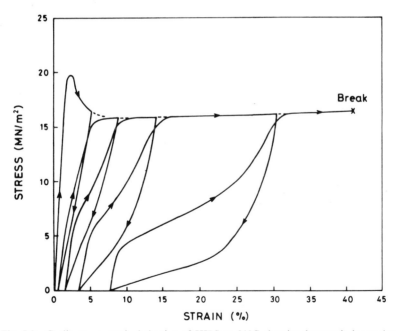

Fig. 5.9 Cyclic stress–strain behavior of HIPS at 21°C showing hysteresis loops due to crazes, which increase in number during the test [Bucknall (22), reproduced by permission].

[21], who studied torsional damping as well (at ~1 Hz) in HIPS, and were able to correlate these two characteristics. At a critical axial tensile stress, damping became dependent on the amplitude; at this point, crazing was observed. With repeated stress–strain cycling, they obtained curves resembling those shown in Fig. 5.9. As might be expected, crazing (and amplitude-dependent damping associated with crazing) was observed at a tensile stress lower than that expected for static tensile loading; thus, crazing was observed at a load of 11 MPa in fatigue, but not in a creep test at the same load over a much longer period of time.

In another study of dynamic mechanical response as a function of cycling (at 12 Hz in tension), Murukami *et al.* [25] described effects of fatigue on the dynamic mechanical properties (at 138 Hz) of nitrile-rubber-modified PVC, PMMA, and PS in comparison with unmodified PS and PMMA. Changes in the dynamic spectrum were small, but there was a slight decrease in the loss modulus E'' for the PS and PVC blends at temperatures less than 0°C. With respect to fatigue endurance at 12 Hz, the presence of the rubber tended to result in a lowering of endurance, especially at $< 10^5$ cycles (see discussion below; improvement is more typical).

The effects of tension–compression cycling between fixed strain limits on unnotched ABS have been described by Beardmore and Rabinowitz [26] (see also Section 2.5). As is the case with other glassy polymers, maximum stresses first dropped rapidly (Fig. 5.10) (corresponding to "strain softening"), and then changed little over about 60 cycles. At this point the tensile stress, but not

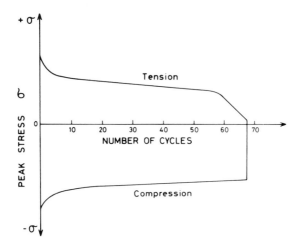

Fig. 5.10   Fatigue of ABS between fixed strain limits at room temperature showing decreases in peak stress with number of cycles [After P. Beardmore and S. Rabinowitz, *Appl. Polym. Symp.* **24,** 25 (1974). Copyright © 1974, by John Wiley & Sons, Inc. Reprinted by permission of John Wiley & Sons, Inc.]

the compressive stress, dropped sharply, and catastrophic failure occurred within a few cycles. These observations are consistent with the development of crazes on the tension half of the strain cycle; during compression, the crazes, being closed up, cannot significantly affect the compliance. The onset of the rapid decrease in tensile stress suggests the onset of craze fracture that leads to a critically sized crack. The changes in maximum stress were also associated with increases in the size of the hysteresis loop, in a manner consistent with that seen in Fig. 5.9.

Menges [27] and Menges and Wiegand [28] studied the fatigue and static behavior (in tension, at 10 Hz) of several resins including ABS and compared dynamic and static creep response in ABS. Menges and Wiegand [28] found the elongations at fracture under fatigue loading (mean stress, 10 MPa; stress amplitude, 8 MPa) to be lower than critical strains for failure under static loading. In view of this finding, the need for caution was urged in designing against fatigue with such materials (see following discussion).

In a recent detailed fatigue study by Bucknall and Stevens [20], specimens of ABS and HIPS were cycled between fixed load limits in fully reversed tension–compression at a very low frequency, 0.03 Hz. Progressive softening and increased hysteresis was observed in each case (Fig. 5.11); significant

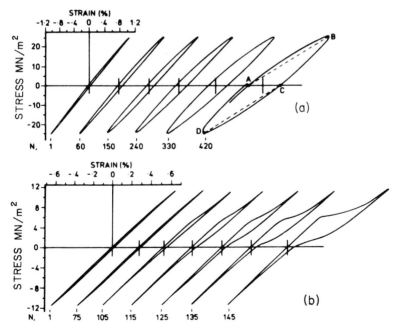

Fig. 5.11 Hysteresis loops at various stages of fatigue in (a): ABS($\sigma = \pm 25.4$ MPa) and (b): HIPS($\sigma = \pm 11.6$ MPa) [Bucknall and Stevens (20)].

differences were observed in the shapes and areas of the hysteresis loops. Thus, with ABS, the tensile and compressive half-cycles exhibited similar energy losses, while with HIPS (after ~ 60 cycles), increasingly higher energy losses were shown in the tensile than in the compressive half-cycle. Moreover, the secant modulus revealed greater softening in tension for HIPS than in compression but no relative difference for ABS. These differences are consistent with the dominance of crazing in HIPS and of shear yielding in ABS (as previously discussed).

As expected, the *average* temperature rise was greater, the higher the amplitude of the applied stress. A typical curve showing temperature as a function of cycling is shown in Fig. 5.12; the oscillations seen in Fig. 5.12 correspond to an alternation between decreasing and increasing temperatures due to the thermoelastic effect. While failure with such loads was ductile (with the internal initiation of cracks), failure at lower stresses (< 27 MPa) was initiated at the surface and was brittle in nature (Fig. 5.13). This figure also compares fatigue fracture with static creep rupture data and shows that, at equivalent time under tensile load, the square-wave loading is more deleterious to material lifetime than a static load. It is also interesting that fatigue failure times can be expressed in terms of stress according to an Eyring-type plot (Fig. 5.14). In Fig. 5.14, the break corresponding to the

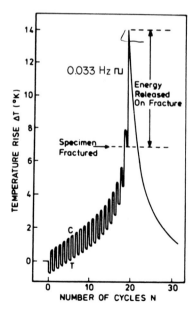

Fig. 5.12 Thermocouple output showing temperature changes during fatigue of ABS at ± 37.5 MPa [Bucknall and Stevens (20)].

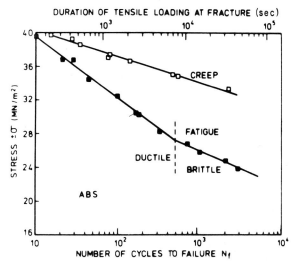

Fig. 5.13 Comparison of S–N curve under square wave loading with creep rupture data for typical ABS [Bucknall and Stevens (20)]. Note that the creep rupture time scale is equivalent to that of the fatigue cycle scale in terms of time under tensile load.

onset of ductile failure in ABS is clearly evident; the displacement of the two ABS curves from each other reflects a difference in rubber content, and the two curves can be made to coincide by normalizing applied stress with respect to yield stress (curve not shown).

All these experiments confirm the importance of considering strain softening and hysteretic heating in specimens that are unnotched or that contain very small flaws. Unfortunately, little evidence exists to permit comparison of fatigue lives in unnotched specimens (by S–N curves) of rubber-modified polymers with unmodified ones. Thorvaldson [29] showed an S–N plot that shows a typical ABS resin to be better than a typical PS, but it is unlikely that the comparison was made keeping the matrix the same. As mentioned above, Murukami *et al.* [25], in tests apparently using equivalent matrix polymers, found a general tendency for rubber additions to decrease fatigue life. Bucknall and Stevens [20] cite problems with fatigue failure in ABS, but a comparison with the equivalent matrix is not given. In any case, as pointed out by Menges and Wiegand [28] and by Bucknall *et al.* [20, 30], caution should be used in designing with rubber-modified plastics that may be expected to encounter high stresses or high strains even at fairly low frequencies.

At the same time, it should be noted that cyclic loads leading to significant heating and gross softening are, in fact, often high relative to the yield stress (see, e.g., Bucknall and Stevens [20]), and may well be outside the range

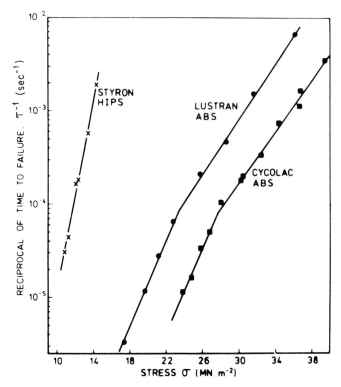

Fig. 5.14 Eyring plots of reciprocal failure time in ABS and HIPS versus stress [Bucknall and Stevens (20)].

permitted by a cautious designer. Since in practice much smaller loads may be anticipated it is important to consider failure at relatively low loads, i.e., failure by propagation of a flaw. Moreover, significant flaws are typically present in fabricated objects.

Fatigue Crack Propagation in Notched Specimens. In general, as is the case with toughness under static loading, the resistance to FCP is increased by the incorporation of a rubbery or toughening phase into a matrix (either brittle or ductile). At the same time, interesting variations in detailed response as a function of frequency and composition may be discerned. As will be shown, these differences probably reflect a balancing of effects due to hysteretic heating (see also Section 3.4).

The general toughening effect has already been illustrated in Fig. 3.30, in which typical commercial rubber-modified polystyrenes, both ABS and HIPS, exhibit lower values of crack growth rate at equal loads than a typical polystyrene [31]. Both ΔK_{max} and (at least for $\Delta K < 2$ for HIPS) $\Delta K^{\neq}$ (the ΔK required to drive the crack at a constant rate per cycle) increases in the

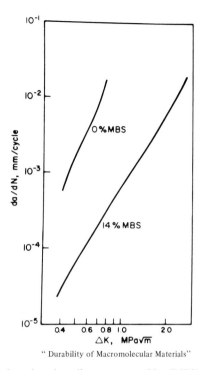

"Durability of Macromolecular Materials"

Fig. 5.15 Effect of methacrylate–butadiene–styrene rubber (MBS) addition (14 phr) on FCP behavior of PVC($M_w = 6.7 \times 10^4$). The curve for 14 phr also represents data for 6 and 10 phr of MBS (phr = parts per hundred of resin) [Skibo *et al.* (33)].

series PS < HIPS < ABS, the same ranking shown by impact strengths (Table 5.1). One may conclude that the differences between HIPS and ABS are due in part to the enhanced dispersion of crazes in the latter (see Sections 5.1.2–5.2.4, inclusive), in addition to the latter's greater propensity for shear response. It is true that this comparison suffers from the fact that the molecular weight of the matrix is unknown and unlikely to be the same in all three materials. Nevertheless, as shown in Figs. 5.15, 5.16, and 3.10, clear evidence for toughening in systems in which M was in fact kept constant is provided by the cases of rubber-modified PVC [32, 33] and toughened* nylon 66 [34, 35]. As expected, at constant rubber content, increasing molecular weight increases fatigue resistance (Fig. 5.16). (The curious shapes and crossovers exhibited by the nylon are discussed in Section 3.4.) In fact, an MBS-toughened PVC is more resistant to fatigue crack propagation than a typical commercial polycarbonate (Fig. 5.17). Hence if one has a specimen containing a significant flaw, the expected fatigue life will be generally enhanced

---

* The precise nature of the toughening phase has not been disclosed.

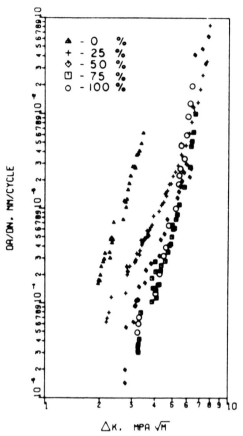

Fig. 5.16   Decreased FCP rates at 1 Hz in blends of nylon 66 with HI-nylon 66: ▲, 0%; +, 25%; ◇, 50%; □, 75%, ○, 100% [Skibo *et al.* (34)].

by the incorporation of a toughening phase (see Section 3.9 for a discussion of prediction of fatigue life from *da/dN* curves).

The occurrence of crazing has been documented in HIPS, ABS, and HIPS-toughened PPO (see Sections 5.1.2 and 5.1.4, and Skibo *et al.* [31]). Fracture surfaces of both the ABS and HIPS/PPO specimens (Fig. 5.18) showed strong whitening, the intensity increasing with increasing ΔK, as expected for crazed material. At low ΔK, bands were observed perpendicular to the direction of crack growth (Fig. 5.18); as noted previously by several investigators for a number of polymers [36–39], the size varied with the square of ΔK (see also Section 4.22). Assuming that the band size corresponded to the size of the crack-tip plastic zone based on the Dugdale formulation, values of the tensile yield (i.e., crazing) stress could be inferred;

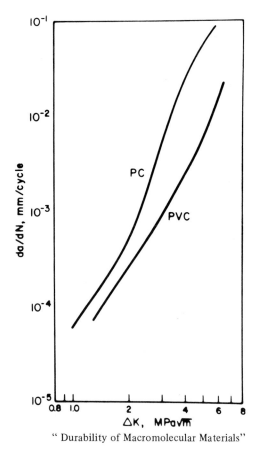

" Durability of Macromolecular Materials"

Fig. 5.17 Comparison of FCP behavior at 10 Hz of MBS-modified PVC ($M_w = 2.08 \times 10^5$); 14 phr of MBS) with that of a typical commercial polycarbonate [Skibo, *et al.* (33)].

values of 43 and 80 MPa were computed for the yield stresses of ABS and HIPS/PPO, respectively [21]. These values are within the range deduced from values of flexural yield stress given for ABS and HIPS/PPO [9]. Thus, with respect to plastic zone size, the two rubber-modified polymers studied behave as do many other plastics.

These bands correspond to the length of the damage zone that has grown to a stable size after many load cycles, and is then fractured in one load cycle; they have been called "discontinuous growth bands" [21] or DGBs (see Section 4.2.2). At higher levels of ΔK ($\Delta K > 1.2$ MPa·m$^{1/2}$) the DGBs disappear, and the crack continuously grows. It is suggested, as before [36], that this transition corresponds to the onset of multiple craze bands parallel to the crack. Figure 5.18 reveals evidence of complex morphologies for both

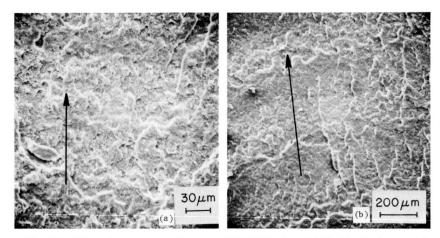

Fig. 5.18 Discontinuous growth bands in HIPS/PPO (a) and ABS (b). Crack growth in direction of arrow [Skibo *et al.* (31)].

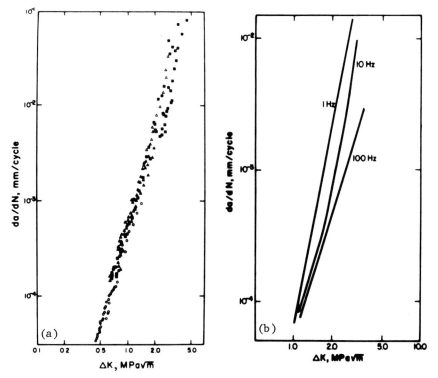

Fig. 5.19 FCP response of ABS (a) ▲, 1 Hz; ■, 10Hz; ○, 100 Hz and HIPS/PPO (b) as a function of frequency [Skibo *et al.* (31)].

ABS and HIPS/PPO. At low ΔK, the surface is rough, with patches of material lying on the surface; the latter probably arise as a result of the crack jumping from one side of the crazed damaged zone to the other [36, 40]. At high ΔK (when crack growth is continuous, see above), there is no evidence in either resin of the classical fatigue striations often observed, perhaps due to the higher than usual roughness. (For a detailed discussion of fractography and the micromechanisms involved see Chapter 4).

While the general toughening effect is well documented, interesting specific effects of frequency and composition may be noted. In fact, the toughened polymers studied so far provide excellent illustrations of the effects discussed above (Sections 2.3.2, 3.4, and 5.1.3). Thus, even though a distinction may often be made between fatigue failure by hysteretic heating (as in the ductile failure on unnotched specimen at moderate frequencies) and failure by propagation of a notch or flaw (nominally "brittle" failure at low frequencies), studies of several toughened plastics reveal that hysteretic effects can also be important in fatigue crack propagation.

First let us examine specific effects of frequency in ABS- and HIPS-modified PPO. As shown in Fig. 5.19, the responses of the two resins are quite different, ABS being relatively insensitive over a broad range of ΔK, and HIPS/PPO being quite sensitive. Also, as shown earlier (Section 3.4), pure PS shows a significant *decrease* in FCP rate at constant ΔK when cyclic frequency was increased (pure PPO having been unavailable for study). Clearly this behavior is not consistent with predictions of the thermal failure mechanism that failure should occur sooner, the higher the frequency. On the other hand, it is useful to consider the possible role of a secondary transition whose characteristic temperature–frequency combination corresponds to the prevailing stress conditions [41, 42]. It may be clearly seen from Fig. 5.20a that the curves of E'' and tan δ versus temperature are nearly flat for ABS at 110 Hz. The peaks at 120 and $-90°C$ correspond to the glass temperatures of the glassy and rubbery phases, respectively. In the case of HIPS/PPO, Fig. 5.20b shows that the damping begins to rise at $\sim 20°C$ and exhibits a shoulder at $\sim 38°C$ on a broader peak centering at $\sim 110°C$. After a slight drop, the damping then rises monotonically until the principal peak is reached at 215°C.

In any case, following the lines of the arguments in Section 5.1.2, if hysteretic heating occurs *at the crack tip*, the temperature rise should be *lower* with the ABS than with HIPS/PPO; the temperature T should reach a steady state, for E'' and tan δ are essentially independent of temperature over a wide range. On the other hand, with HIPS/PPO, temperature will increase in an autoaccelerating manner, for E'' and tan δ themselves increase with temperature. Further, the magnitude of E'' (and tan δ) in HIPS/PPO is twice that in ABS at room temperature. Both of these effects imply an increase in

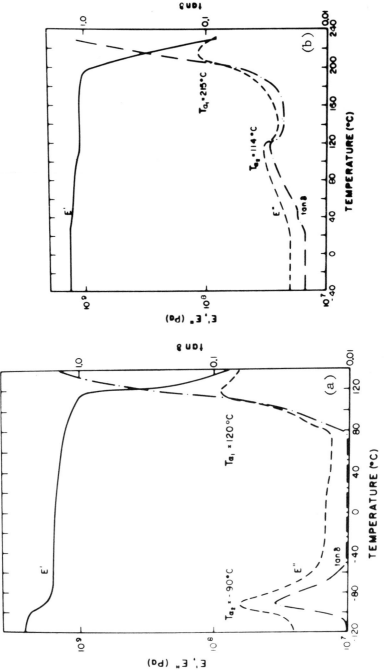

Fig. 5.20 Dynamic mechanical response of ABS (a) and HIPS/PPO (b) as a function of temperature (frequency = 110 Hz) [Skibo et al. (31)].

the rate of hysteretic heating in HIPS/PPO relative to ABS. Indeed, sub-stantial increases in temperature at the tips of growing cracks have been observed [34, 35, 44]. Such temperature rises in notched specimens do *not* necessarily imply a thermal failure mechanism, for otherwise the overall failure should be faster at higher frequencies. However, as suggested earlier [41], *highly localized* hysteretic heating may result in localized yielding and an increase in the crack-tip radius. Since ΔK is an inverse function of crack-tip radius, the *effective* ΔK should be lowered; in consequence, da/dN should also be lowered at the *nominal* ΔK applied. Thus, the frequency response of at least these two rubber-modified resins (ABS and HIPS/PPO) is consistent with the response of typical unmodified polymers, in that high sensitivity to frequency is associated with the existence of a secondary transition close to the frequency and temperature of testing (see Section 3.4). The point may be not that the transition corresponds to a secondary relaxation process, but that a transition exists that can result in an elevation of E'' and tan δ, with consequent localized heating of the crack tip.

Interesting effects of composition and M are evident in MBS/PVC [32, 33]; as with the effects of frequency just discussed, hysteretic heating (and especially the size of the volume affected) appears to be a principal factor. As mentioned above, toughness was clearly increased by the incorporation of MBS (Figs. 3.30a, 5.17, 5.21, and 5.22). In contrast to the case of impact behavior, in which improvement needed from 6 to 10 phr rubber, only 6 phr rubber was needed to increase ΔK_{max} several fold. However, it may be noted that values of ΔK_{max} could be estimated only for specimens having low $M(M_w \leq 1 \times 10^5)$ and low rubber content (6 phr). At higher values, failure became ductile, presumably due to the onset of significant hysteretic heating over a relatively large volume as ΔK increased to high levels, even though only a slight rise in temperature at the crack tip was observed. The resistance to fatigue crack growth was also increased (Figs. 3.30a, 5.17, and 5.21) by the incorporation of as little as 6 phr MBS. Indeed, for the lowest-M material, the value of da/dN at a value of $\Delta K = 0.9\Delta K_{max}$ (of the control) was reduced *40-fold*! On the other hand, with the highest-M matrix, da/dN was reduced to a relatively lesser extent, only fivefold (the curve in Fig. 5.21 also holds for 6 phr MBS). At the same time, in the latter case, the failure became ductile in nature. Thus in contrast to the case of impact strength, even a low concentration of MBS (6 phr) induces a significant toughening in fatigue even at a low value of M ($M_w \sim 7 \times 10^4$).

With both the lowest-M and the highest-M matrixes, the addition of >6 phr of MBS had no significant effect on FCP rates. This behavior might speculatively be attributed to the consequences of a dynamic balance between (1) rubber-induced toughness due to the ability of the rubber to induce shear yielding and crazing and (2) rubber-induced softening (that is, reduction in

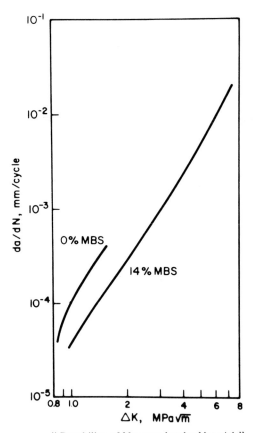

"Durability of Macromolecular Materials"

Fig. 5.21   Effect of MBS addition (14 phr) on FCP behavior of PVC ($M_w = 2.08 \times 10^5$). The curve for 14 phr also represents data for 6 and 10 phr of MBS [Skibo *et al.* (33)].

stiffness). Now stiffening can be reduced by two factors, acting alone or together: the presence of the rubber *per se* (see, for example, Fig. 5.6), and hysteretic heating due to inherent mechanical damping. It may be recalled that, depending on the system, temperature, and presumably frequency, E'' or tan δ may increase, decrease or remain the same as the modifier content is increased. It is interesting that the modified PVCs, but not the controls, tended to heat up to some degree during testing. Any such heating could have several possible and sometimes contradictory effects, for example, increased creep and blunting of the crack tip (lower FCP rates), and reduced modulus E (adding to the reduction due to the presence of the modifier, thus increasing FCP rates). One might also speculate that when M is high, the

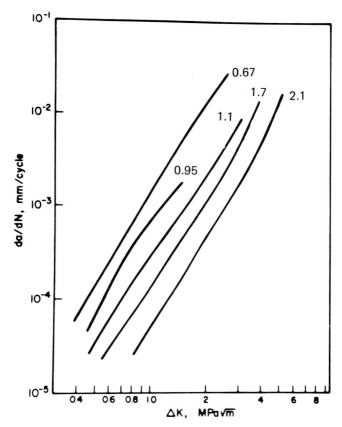

Fig. 5.22 Effect of molecular weight on FCP behavior of PVC ($M_w \times 10^5$) modified with 6 phr of MBS [Skibo *et al.* (33)].

toughening effect due to crack blunting or to the stimulation of dissipation in shear is *relatively* less prominent than in a lower-*M* material.

In any case, in an absolute sense, increases in both molecular weight and MBS content are highly beneficial to fatigue crack propagation. One may conclude that by combining the inherent ability of a higher-*M* and ductile matrix to dissipate strain energy with the enhancement of highly delocalized deformations by the rubber particles [probably mainly in shear at the low strain rates involved here (see Bucknall [7, Chap. 8])], one can obtain a much greater level of fatigue crack resistance than is possible with, for example, rubber-toughened polystyrene. As already mentioned, toughening by addition of a rubbery phase is most effective in a ductile polymer; fatigue behavior also reflects this observation.

Still more complex effects of frequency and composition are possible. It is illuminating to consider the striking effects of both frequency and composition in toughened nylon 66 [33, 34, 43] and the role of hysteretic heating on various scales. While the presence of the toughening phase resulted in a general improvement in FCP resistance (Fig. 3.10), several anomalies in fatigue response are evident (see Section 3.4). Until this study was completed, the increasing frequency had always either *decreased* FCP rates (as with HIPS/PPO) or had essentially no effect (as in ABS or nylon 66), observations inconsistent with arguments for a thermal failure mechanism. However, with impact-toughened nylon 66, increasing frequency resulted in an *increase* in FCP rates (Fig. 3.11b) at constant ΔK. Since, in fact, temperatures as high as 125°C were measured close to the tip of the growing crack in moisture-equilibrated material, a major effect of hysteretic heating is implied (recall Fig. 3.11).

In summary, the fatigue response of toughened plastics is dominated by the characteristically higher level of viscoelastic damping relative to that of the unmodified matrix. When the volume of material experiencing the cyclic load is large (as in unnotched specimens, especially at high frequencies and loads), failure tends to occur by gross softening due to hysteretic heating. When the damaged zone is small (as in notched specimens of some materials), localized heating may be beneficial in blunting the crack tip. With notched specimens of still other materials, the damaged zone may become large enough that gross heating and softening of the matrix may enhance crack propagation. Thus we have an interesting range of behavior between the extremes of failure by gross heating and failure by propagation of a crack or flaw.

5.2. PARTICULATE-FILLED COMPOSITES

5.2.1. Introduction

Since polymers filled with rigid phases are frequently used in engineering as well as other applications, their stress–strain behavior has received much attention, and principles of fracture mechanics are often used to characterize their toughness. (For reviews and discussion, see Manson and Sperling [3, Chap. 12]; Krokosky [45], and Bucknall [46]; as mentioned above, discussion excludes rubbery matrixes.) Most commonly inorganic fillers such as silica, glass or mica are used [47], but preformed rigid polymer particles may also be incorporated, as in the casting *in situ* of a rigid-polymer-filled resin for use in orthopedic or dental restoration [48–57]. If the concentration of the filler is high enough that the polymer serves mainly as a binder, the resultant composite is often called a "polymer concrete" (see Section 5.5).

5.2.2   Fatigue Behavior

Curiously, little has been reported on fatigue in particulate-filled composites based on rigid matrixes, and most that has been published involves dental or bone cements.

At least in the English literature* one of the first studies of fatigue in plastics filled with particulate inorganic materials appears to have been reported by Nielsen [60], who studied the fatigue life of nylon 6 filled with glass beads or silica, nylon 66 filled with a clay, and an epoxy filled with glass beads. Tests were run by the rotating beam method. In general, dynamic mechanical response remained nearly constant until just before failure. Hysteretic heating was proposed to explain the decrease in modulus as failure was approached. Unfortunately, the preliminary data presented do not permit further analysis.

With typical restorative materials, a polymeric powder, usually at least partially acrylic in nature, is mixed with pigments or fillers, a small proportion of acrylic monomer (which may contain a cross-linking agent), and a free-radical initiator [48]; after kneading and application, the mixture is then polymerized. It is frequently reported that the tensile strength and fracture toughness leaves much to be desired, these properties often having values lower than for a typical glassy polymer or at best somewhat higher [49]. One problem is that the toughness of such materials depends strongly on the fabrication technique, which is difficult to optimize under normal working conditions. These difficulties are important, for fatigue resistance in dental composites is believed to be correlated with fracture toughness [50, 53] (see also Section 3.8.7); the relatively low values of toughness (and fatigue resistance) are undoubtedly partially responsible for the relatively short service lives of dental composites (1–4 yr). An analogous problem may well exist for bone cements.

Apart from the study of fracture toughness *per se*, one study of fatigue crack growth has been reported by Lankford *et al.* [52]. As shown in Fig. 5.23, it was found that the fatigue crack propagation response obeys the Paris relationship [Eq. (3.10)]. Also, the microstructure of the cured composites affected the slopes of the FCP curves; the specimens kneaded to minimize microvoids were somewhat less resistant to crack growth than were those kneaded in the presence of whole blood (the liquid phase contributing the effect of a toughening microvoid phase). It was also found that all composites were more resistant to crack growth (at least at $\Delta K > \sim 3$–$4\ \text{MPa} \cdot \text{m}^{1/2}$) than typical samples of PMMA. These findings were rationalized after fracto-

---

* There are reports of other studies published in other languages (see, for example, Aoki [58] and [59]); however, translations were not available at the time of writing.

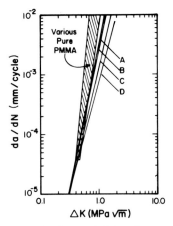

Fig. 5.23   Fatigue crack propagation in a typical dental cement, Simplex P. [After J. Lankford, W. J. Astleford, and M. A. Asher, *J. Mater. Sci.* **11**, 1624 (1976), Chapman & Hall, Ltd.] A, B, C, D are dental cement, with C and D containing microvoids.

graphic examination, which revealed a combination of several crack-resisting mechanisms: pinning of the crack front by the original solid particles, lower crack growth rates within the matrix, and diversion of the crack tip through the nucleation of microcracks at void sites. The lower crack growth rate in the matrix was ascribed to incomplete polymerization (presumably on the assumption that plasticization by monomer should increase toughness). Results were consistent with the findings of Beaumont and Young [49], who studied crack propagation in the same material under static loading.

The mechanical properties, including fatigue response, of two acrylic bone cements have been studied by Freitag and Cannon [55, 56] as a function of composition, mixing technique, and environment (the latter included tests in bovine serum at 37°C to simulate *in vitro* conditions). Tests were conducted in the rotating–bending mode at frequencies in the range 20 to 23 Hz; at these frequencies failure was by crack growth rather than softening. Examination of the S–N curves obtained indicated that fatigue life was greater in the bovine serum than in air at ambient temperature. With one cement, $BaSO_4$ (10 wt%) was incorporated as a radio-opaque filler. The $BaSO_4$ filler increased the fatigue life in air, but had no significant effect on the lifetimes in bovine serum. No significant differences were found in the fatigue responses of the two cements [55–57]. Pilliar *et al.* [61, 62] also studied the fracture toughness and fatigue resistance of bone cements, some containing carbon fibers; their work is discussed in more detail in Section 5.3.

Clearly much research is still needed with particulate-filled composites. As has been pointed out by several investigators earlier (see Manson and

Sperling [3, Ch. 12] and Koblitz *et al.* [53]), fracture properties depend in a very complex manner on the interaction between the filler and the matrix, and many possible toughening mechanisms exist. The proprietary nature of the formulations, and the need to bond to a complex substrate, also make it difficult to do controlled studies. However, some promising attempts to improve behavior by modifying interfacial bonds have been reported [51, 59]. The concept that an ionomeric bond at an interface should be advantageous is quite consistent with current ideas of acid–base interaction between constituents in a composite [63, 64; see also Section 5.5 and Fowkes and Mostafa [64]. Studies of fatigue as well as static response in terms of inter-facial properties should be of fundamental and practical importance. (To facilitate such studies, at least with dental materials, Koblitz *et al.* have designed a special test specimen and procedure [53].)

5.3 FIBROUS COMPOSITES

As the shape of a high-modulus dispersed phase is elongated from a sphere to a fiber, the filler stiffens the composite to a greater extent (at a given volume fraction) and begins to carry an increasing fraction of the load [65–67]. While glass fibers have long been used in composites, other fibers are also of contemporary interest, e.g., boron, carbon,* and highly oriented organic fibers; hybrids are often advantageous, e.g., glass–boron systems. A wide range of configurations and compositions is used, from composites containing a few volume percent of randomly dispersed, short discontinuous fibers to unidirectional or cross-plied arrays of continuous fibers bonded together by a polymer.

With short-fiber composites (Section 5.3.1), the reinforcement typically yields a material whose mechanical properties are superior to those of the matrix. With continuous-fiber systems (Section 5.3.2), in which the fibers constitute most of the composition, the bonding matrix protects the fibers and makes possible better utilization of the inherently high levels of mechanical properties of the fibers than would otherwise be the case. The importance of the matrix (and of the fiber–matrix bond) is obvious in the case of short-fiber composites but must also be kept in mind with continuous-fiber systems. Indeed, in many combinations of loading and composite geometry, the matrix is a strength-limiting variable, for example, when fibers are aligned at an angle to the axis of loading, and when shear stresses are involved.

---

* In fibers based on carbon, the carbon is graphitized. While both "carbon" and "graphite" are used to refer to these fibers, the term "carbon" will usually be used in this monograph.

Since continuous-fiber composites have been extensively used in applications involving cyclic loads, a voluminous literature on fatigue of such systems has developed. Although less has been published on short-fiber systems, considerable information on their fatigue behavior is now available; for example, the subject of fatigue in composites has been reviewed by Dew-Hughes and Way [68] and Owen [69]. More recently, Harris [70] has summarized the effects of fatigue loading and outlined problems in understanding the mechanisms involved; also Reifsnider [71] has critically analyzed the phenomena of fatigue in composites. Several conferences have been held, some devoted entirely to fatigue of composites [72–75] and some to related topics [76–83]. Discussions of fatigue and fracture generally may also be found in references [2, 67 and 84–94], inclusive, as well as in the general literature.*

Much of the research reported is experimental and empirical, the purpose being to develop information useful to the designer of structures. In this section the principal features of fatigue in fibrous composites are outlined and specific examples of problems and approaches to fatigue are discussed. Although a detailed treatment of all systems is beyond the scope of this book, the references cited should help the reader who seeks specific information.

Before beginning discussion, it will be useful to enumerate some of the major complications in the study of fibrous composites, some in common with nonpolymeric composites, and some peculiar to polymeric systems:

(1)   Criteria for failure vary widely; while damage is progressive, physical integrity may be maintained for many decades of cycles [87].

(2)   Many individual processes are involved in the failure of fibers, matrix, and interfacial regions (see Table 5.2 and the Appendix); processes involving the polymer are subject to variations in state due to hysteretic heating and time-dependent behavior.

(3)   Thus the nature of damage is complex [71], depending on the mode of loading, the relative orientation of the stress and fiber axes, the coupling of stress fields, constraints due to in-plane and through-thickness elements in continuous-fiber systems and the presence and nature of flaws.

(4)   The concept of a crack needs redefinition, and, at least for some purposes, the conventional fracture mechanics approach may also need revision [71].

(5)   While strength and fracture toughness generally decrease on cycling, the occurrence of certain combinations of individual failure processes with

* Especially good sources are the International Journal of Fracture, the Journal of Composite Materials, the Journal of Materials Science, Composites, Polymer Engineering and Science, and proceedings of the meetings of the Reinforced Plastics/Composites Institute of the Society of the Plastics Industry (New York). Conferences and symposia sponsored by the Society of Plastics Engineers, the American Chemical Society, the Metallurgical Society (AIME), and the ASTM are often useful; the conferences on deformation and fracture in polymers sponsored periodically by the Plastics and Rubber Institute (London) are especially relevant.

TABLE 5.2 Factors Contributing to the Work of Fracture[a]

Type of work	Symbol	Origin	Form of energy dissipation
Fiber internal work	γ_{fb}	Fiber brittle fracture	Stored elastic energy
	γ_{fs}	Fiber bending during pullout	Plastic flow during bending
	γ_{fd}	Fiber ductile fracture	Plastic flow and necking
Interface work	γ_{mf}	Difference in tensile strains across interface	Frictional sliding or plastic shear in matrix
	γ_{fp}	Fiber pullout	Frictional sliding or plastic shear in matrix
Interface and matrix work	γ_{ms}	Splitting of matrix parallel to fibers	Matrix surface energy and fiber–matrix bond energy
Matrix internal work	γ_m	Matrix fracture	Matrix surface energy and plastic flow

[a] Based on Piggott [93], with permission from Chapman & Hall Ltd.

certain fiber arrays may result in effective crack blunting so that fracture toughness may *increase*, at least over part of the fatigue life [71, 92].

With these points in mind, let us examine typical studies of fatigue in fibrous systems.

5.3.1 Discontinuous-Fiber Composites

The combination of short discontinuous fibers (from, say, 3 to 30 mm in length) with thermoplastic or thermosetting matrixes are of increasing technological interest because the resultant materials can be processed by injection molding, compression molding, or extrusion techniques generally similar to those used for polymers themselves [95]. The most common fibrous reinforcements* are glass, graphitized carbon, and acicular minerals

---

* The term "reinforcement" means different things to different people. In plastic-based composites, the term usually refers to the improvement of modulus, and sometimes strength. With fibrous composites containing high-modulus fibers, the modulus *and* strength are normally increased in proportion to the volume fraction of fiber. With particulate composites, based on high-modulus fillers, the modulus, but not always the strength, is increased relative to that of the matrix (3, Chapter 12).

It should also be noted that the comparison of properties requires the use of *volume* not *weight* fraction of reinforcement. Thus, for example, at an equivalent weight fraction, carbon fibers, which have a lower density than glass, will have a higher volume fraction than glass. This fact undoubtedly plays a significant role in the superiority of carbon fibers as a reinforcement for plastics.

such as asbestos (for which today noncarcinogenic replacements are being sought); sometimes glass/carbon hybrids are used as well. A wide range of fiber concentrations and geometrical arrays is used, depending on the application. One common type is a composite containing from ~ 2 to $\sim 40\%$ (by weight) of fibers that can be injection-molded; considerable orientation in the flow direction may be expected with this type. A second common type is based on the impregnation of a more or less isotropic mat that is made by laying up short fibers (typically, chopped strands) in a random manner, impregnating the mat with a prepolymer, and curing the matrix. The mechanical response in such a case will be essentially isotropic, but the properties in a given direction will be lower than in the case of preferential fiber orientation along the stress axis. Of course, chopped strands can be laid up to yield anistropic specimens as well and short fibers are often blended with long fibers and particulate fillers [95]. In fact, the combination of low cost with good strength, stiffness, fatigue resistance, and dimensional stability of short-fiber-reinforced plastics, especially relative to their density, has led to a major penetration into markets previously held by metals [96].

General Aspects of Fatigue in Discontinuous-Fiber Composites.   It is true that maximum improvements in the stiffness and strength of fibrous composites are attained with continuous fibers tested in the fiber direction [66]. However, for length/diameter (l/d) ratios greater than about 60, even short fibers can be loaded to their maximum possible stress if uniaxially oriented. Since in practice l/d ratios are usually > 150, such fibers can resemble continuous ones in their ability to carry load transferred from the matrix. However, in many cases less anisotropy is common, for deliberate or fortuitous reasons. Obviously properties will depend on the average degree of orientation of the fibers with respect to the loading axis. As the proportion of fibers having angles $> 0°$ with respect to the loading axis increases, and as the angle increases, the failure properties will be increasingly dominated by the matrix and interfacial properties.

The importance of aspect ratio to fatigue performance and the major role of the matrix in composites containing off-axis fibers (with respect to the stress axis) has been well illustrated by Lavengood and Gulbransen [97]. Working with short boron fibers (50–50 vol%) embedded in an epoxy resin, they studied the effect of cyclic loading on the number of cycles required to produce a 20% decrease in elastic modulus, at a frequency of 3 Hz (low enough to minimize hysteretic heating). While the fatigue life (as just defined) increased with the aspect ratio up to a value of $l/d = 200$ (see Fig. 5.24), further increases in l/d had little effect. This is fortunate in view of the fact that fabrication becomes difficult with longer fibers.

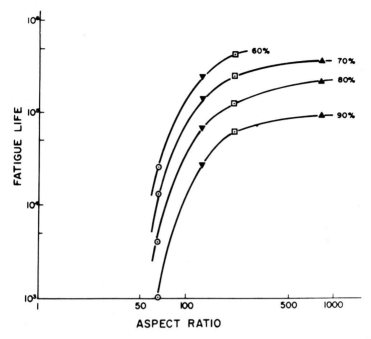

Fig. 5.24 Cycles to failure vs aspect ratio for an applied stress at 60, 70, 80, and 90% of ultimate failure stress [Lavengood and Gulbransen (97)]. Boron-fiber-reinforced epoxy.

In all cases, the failure mechanism was found to be a combination of interfacial fracture and brittle rupture of the matrix at 45° to the fiber axis.

At higher frequencies, one may expect hysteretic heating (see Section 2.3.2) to become a significant factor in fatigue performance. A rise in temperature during cycling may have two effects: a reduction in Young's modulus, and a concomitant decrease in the *effective* fiber length and hence an increase in the critical l/d ratio [98, 99]. The deleterious effect of hysteretic heating on fatigue life has been shown not only for polymers themselves, but also for several short-fiber composites by Cessna *et al.* [99]. At the same time, there is some evidence that fibers may help minimize hysteretic heating, at least in some cases.

Thus, in experiments with flexural fatigue at fixed stress limits of oriented glass-fiber-reinforced nylon 66, Bucknall *et al.* [30] showed that at least at a low frequency (0.5 Hz), the fibers essentially eliminated the dominance of thermal softening in the high-load region so that the portion of the S–N curve at $N < 10^5$ was shifted significantly in the direction of higher stresses (see Fig. 5.25). The difference between the curves for the two fiber orienta-

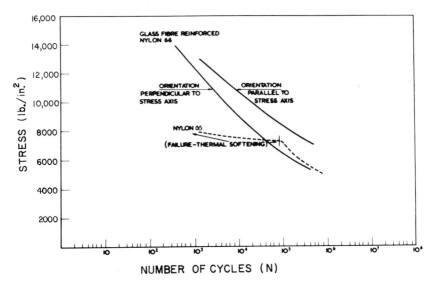

Fig. 5.25 Flexural fatigue of nylon 66 and a glass-fiber-reinforced nylon 66 in dry air at 20°C, at 0.5 Hz, with mean stress = 0 [Bucknall *et al.* (30)].

tions with respect to the stress axis presumably reflects the lower modulus perceived with loading transverse to the fiber axis. On the other hand, the benefits of fibrous reinforcement disappeared when the S–N curve was plotted in terms of strain rather than stress; in fact, in terms of *strain* the matrix *per se* appeared superior to its composite. Since such a clear difference in response exists, one should be careful to match a material to its anticipated cyclic loading conditions.

As mentioned above, the matrix must play a dominant role in all short-fiber composites [100]. This expectation was confirmed in a study by Dally and Carrillo [101], who characterized the fatigue behavior of glass-reinforced nylon, polystyrene, and polyethylene. Although in all cases failure began with debonding at the interface, the composites differed greatly in terms of subsequent growth of cracks into the matrix. Thus cracks propagated readily into polystyrene but not into nylon or polyethylene. Similarly, Soldatos and Burhans [102] reported that toughened epoxies led to better fatigue lives in glass-fabric-reinforced laminates; while Owen and Rose [103] found that flexibilizing a polyester/glass–fabric laminate did not improve fatigue life, the fracture toughness was not changed much either. These findings are, of course, consistent with the general ranking of polymers with respect to FCP resistance and its correlation with fracture toughness (see Section 3.8.7). The interfacial strength must also be important to fatigue resistance. As pointed out by Owen [69], there will always be tensile interfacial stresses somewhere in the composite; test temperatures above or below the matrix cure temperature will inevitably result in internal tensile and compressive stresses. Although the optimum value for interfacial strength is not known, Cessna *et al.* [99] found that use of a coupling agent on glass fibers markedly improved fatigue life.

More detailed discussion of specific effects of fatigue on viscoelastic response, fatigue life, and fatigue crack propagation follows.

Effects of Fatigue on Viscoelastic Response. Kodama [104, 105] has described the effect of cyclic loading (at 10 Hz) on the dynamic mechanical response of unnotched cross-linked polystyrene and unnotched cross-linked poly(methyl methacrylate) as a function of cycling temperature (15, 45, and 65° or 70°C) and glass-fiber content (0 and 5%, by volume) in short-fiber-reinforced (random-in-plane) polymers. While no significant effect of cycling on the matrix alone was observed with either unmodified polymer (after 3.5×10^5 cycles), major effects were noted with all the composites. With the reinforced PS system (SC series in Fig. 5.26a), an additional dispersion peak in the modulus E'' may be seen on the high-temperature side of the α-peak (corresponding to the T_g). This peak was resolved into two components; one was attributed to the development of a spectrum of relaxation

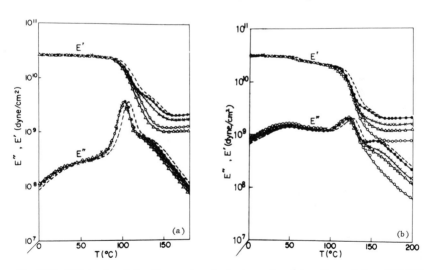

Fig. 5.26 Effects of cyclic loading on viscoelastic properties of glass-fiber-reinforced PS (SC) (a): ●, SC; ×, SC-F$_{15}$; ○, SC-F$_{45}$; △, SC-F$_{65}$; ---, SC (100Hz) and PMMA (MC) (b): ●, MC; ×, MC-F$_{15}$; ○, MC-F$_{45}$; △, MC-F$_{70}$; ---, MC (100 Hz). The letter F refers to specimens fatigued at temperatures corresponding to the numerical subscripts (in °C). [M. Kodama, *J. Appl. Polym. Sci.* **20**, 2853 (1976). Copyright © 1976, by John Wiley & Sons, Inc. Reprinted by permission of John Wiley & Sons, Inc.]

times associated with restraints on the mobility of the polymer chains in volume elements at fiber intersections (see the discussion on interfacial mobility [3, Ch. 12], and the other to interfacial interaction along the fiber sections between crossovers. As the temperature of cycling was increased, two effects were observed: a progressive decrease in rubbery modulus (E'_r, the value of E' at a constant temperature above $\sim 105°C$), and a progressive weakening of the additional damping peak associated with the fiber–matrix interaction. Similar decreases in E'_r may be seen for PMMA in Fig. 5.26b, though the relative effect of temperature is not as consistent. Interestingly, the additional damping peaks, and hence the changes in them, are, however, more pronounced with PMMA than with PS. In any event, cycling appears to erase the thermal history associated with molding at a high temperature followed by cooling.

In effect, Kodama proposes that cycling effectively damages both the interstitial volume elements at the fiber intersections and the interfacial region (or "interphase") itself. To account for the difference in behavior between PMMA and PS, Kodama suggests that the thermal stresses at the fiber intersections are greater for PMMA than for PS. Since fabrication temperatures were similar, one might expect the thermal stresses and the associated restrained volume elements of the matrix to be similar. An explanation may be suggested, however, based on recent observations of

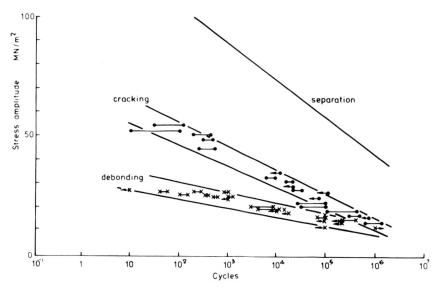

Fig. 5.27 S–N diagram showing the various stages of failure in a chopped–strand–mat/poly-ester resin laminate: fully reversed stress, 1.67 Hz, 20°C, and 40–42% relative humidity. [From Owen *et al.* (108).]

filler–matrix interactions in polymer-impregnated concrete (PIC) and silica-matrix systems [62, 63, 106]. These authors have shown that adsorption at a siliceous (or other) substrate is often associated with the occurrence of strong charge–transfer interactions; thus silica or silicates contain surface acid sites that can enhance the adsorption of a basic polymer such as PMMA in preference to a less basic one. Indeed with PIC systems, PMMA gives significant evidence of restricted mobility (as shown, for example, by a higher than usual T_g) while PS does not (see also Section 5.4). Further studies of interactions such as those implied by the Kodama work would be helpful.

Fatigue Life: General S–N Response. Owen has reviewed major aspects of tensile fatigue damage in glass-fiber-reinforced systems [69]. In an early study by Owen and Dukes [107] using a chopped-strand-mat/polyester laminate, the sequence of events during failure was observed to consist of debonding, cracking in the matrix, and final separation. Subsequently [108], it was possible to develop S–N curves that revealed these stages of failure (Fig. 5.27). Figure 5.27 also shows that damage occurs at quite low fractions of the static strength, and exemplifies the problem with defining failure in terms of phenomenological damage. In any event, fibers lying transversely with respect to the stress axis served as sites for the initiation of debonding.

In general (Figs. 5.28 and 5.29), fibrous reinforcements tend to improve the fatigue strength of a polymeric matrix. In view of the fact that carbon fibers are typically at least three times as stiff as glass fibers and ten times as thermally conductive [69], it is not surprising that fatigue resistance is enhanced more with carbon than with glass fibers (Figs. 5.28–5.30), though, as mentioned, each type tends to strengthen a variety of matrix resins [109, 110]. As implied above, one reason for the superiority of carbon to glass may be due to its much higher thermal conductivity, which should thus lessen hysteretic

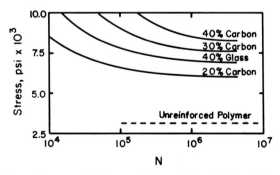

Fig. 5.28 Fatigue endurance of carbon- and glass-reinforced nylon 66 (percentages of fiber given by weight). [Reprinted with permission from J. Theberge, B. Arkles, and R. Robinson, *Ind. Eng. Chem., Prod. Res. Dev.* **15,** 100 (1976). Copyright by the American Chemical Society.]

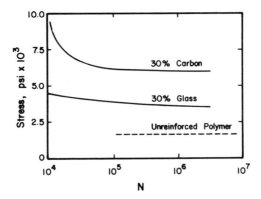

Fig. 5.29 Fatigue endurance of carbon- and glass-reinforced ethylene-tetrafluoroethylene copolymer (percentage fiber given by weight). [Reprinted with permission from J. Theberge, B. Arkles, and R. Robinson, *Ind. Eng. Chem., Prod. Res. Dev.* **15,** 100 (1976). Copyright by the American Chemical Society.]

heating at a given frequency and load. Interestingly, the fatigue endurance limit of 9500 psi (66 MPa) for the 30%-carbon-fiber-reinforced polyphenylene sulfide was stated to be the highest value observed for any reinforced thermoplastic molding compound.

Carbon fibers are also of interest because of their compatibility with living tissue, an important point in prosthetic materials (see Section 5.2.2), which also can benefit from improvements in static and fatigue strength. To investigate the possible benefits of carbon fibers in such materials, Pilliar *et al.* [61] incorporated 2 vol% of carbon fibers (nominally in a random array) in a

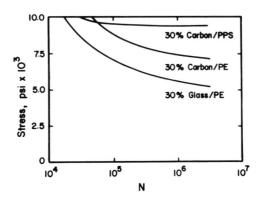

Fig. 5.30 Fatigue endurance of carbon- and glass-reinforced poly(phenylene sulfide) and thermoplastic polyester (percentage fiber given by weight). [Reprinted with permission from J. Theberge, B. Arkles, and R. Robinson, *Ind. Eng. Chem., Prod. Res. Dev.* **15,** 100 (1976). Copyright by the American Chemical Society.]

commercial bone cement and tested the composites with both fatigue and static loading. Fatigue tests were conducted in the tension–tension mode at 1 Hz, a frequency selected to simulate anticipated in-service loading. As well as improving the tensile strength, Young's modulus, and notched and unnotched impact strength (by factors of 1.6, 2.0, 5.4, and 2.2, respectively), carbon fibers, though present at a low concentration, greatly improved the fatigue performance (Fig. 5.31a). As was reported above for a glass-fiber-reinforced nylon, the lifetime was controlled by the strain (Fig. 5.31b), but in contrast to the case of the nylon composite of Ref. [30], data for both the matrix and the composite fell on the same curve. In any case, at a given load, the fatigue-sensitive matrix will undergo a smaller strain in the presence of the fibers. At the time of writing of the paper mentioned, clinical tests had been in progress for 18 months, with no adverse effects noted. It should be noted that the physiological environment of the body is hostile to many materials, including polymers, and may lead to environmentally accelerated fatigue failure.

Continued research by Pilliar *et al.* [62] confirmed a significant (50%) increase in fatigue strength (stress control, 1 Hz) of their carbon-reinforced acrylic bone cement (see Section 5.2.2) in comparison to the unreinforced material at 10^5 cycles (15 versus 10 MPa, respectively). Not surprisingly, the scatter band of the S–N curves for specimens made by different techniques and containing 1 and 2% fiber by volume was broad. (The fatigue strengths of the bone cements were still lower than that of monolithic PMMA, which is not in itself very fatigue-resistant; thus there is still room for improvement.) In fact, the more porous composites obtained under operating-room conditions for fabrication were superior in the high-cycle range to more dense specimens (see also Section 5.2.2); as shown by SEM, the fibers in the latter tended to occur in bundles, which undoubtedly played a major role as initiators of failure. There may also be an effect of environment; a few data suggested that testing in Ringer's solution at 37°C (to simulate body conditions) gave somewhat improved fatigue life.

The effect of notches on fatigue behavior in short-glass-fiber reinforced polymers has received some attention. Misaki and Kishi [111] studied the effects of sharp and blunt notches (the blunt one having a 1.5-mm-radius rounded tip at the base of a 60° notch) on the tensile and fatigue behavior (at 10 Hz) of polycarbonate itself and polycarbonate containing 0.3 vol% short glass fibers. As expected, both types of notch lowered the static tensile strength, to about 40 and 90% of the original value for sharp and dull notches, respectively (see Table 5.3); smaller decreases were noted for the composites. However, much greater decreases in strength were found in fatigue, and the failure mode was brittle rather than ductile as in the case with the unreinforced PC with the dull notch. Thus, while with some composites cycling may

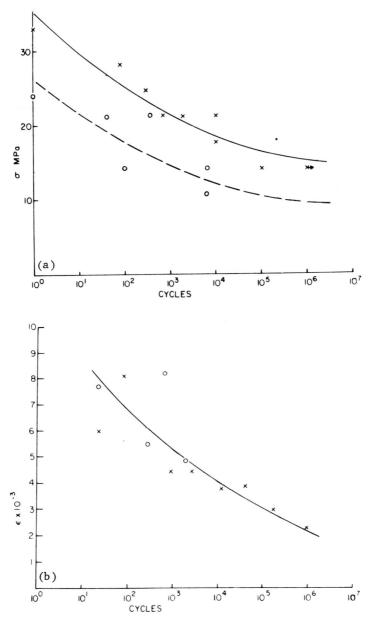

Fig. 5.31 Fatigue-test results (S–N curve) for (×) carbon-fiber-reinforced (2 vol%) and (○) unreinforced normal bone cements. Curves in (a) as a function of stress; curve in (b) as a function of strain. [R. M. Pilliar, R. Blackwell, I. Macnab, and H. U. Cameron, *J. Biomed. Mater. Res.* **10**, 893 (1976). Copyright © 1976, by John Wiley & Sons, Inc. Reprinted by permission of John Wiley & Sons, Inc.]

TABLE 5.3 Effect of Notch Shape on Tensile and Fatigue Strength of
Glass-reinforced Polycarbonate[a]

Material	Tensile strength (kg/mm^2)[b]		
	Unnotched	Sharp notch	Blunt notch
Polycarbonate	6.7	2.7 (0.40)[c]	6.1 (0.91)
Reinforced polycarbonate	10.5	3.6 (0.34)	4.4 (0.42)
	Fatigue strength[d] (kg/mm^2)		
	Unnotched	Sharp notch	Blunt notch
Polycarbonate	5.4	1.1 (0.20)	2.3 (0.43)
Reinforced polycarbonate	7.7	1.4 (0.18)	2.3 (0.30)

[a] From T. Misaka and J. Kishi, *J. Appl. Polym. Sci.* **22**, 2063 (1978). Copyright © 1978,
by John Wiley & Sons, Inc. Reprinted by permission of John Wiley & Sons, Inc.

[b] Conversion: 1 kg/mm^2 = 9.8 MPa.

[c] Values in parentheses indicate the ratio of the strength of the specimen to that of
the unnotched specimen.

[d] After 10^4 cycles at 10 Hz.

actually *increase* fracture toughness [92], this is usually not so with a typical
short-fiber-reinforced system, presumably because the latter lacks the con-
tinuity of fiber that permits extensive crack blunting by fiber reorientation or
delamination in the damaged zone of a continuous composite.

Significant differences in fracture mechanisms were observed by DiBenedetto *et al.* [112] in their statistical study of fatigue failure in graphite-fiber-
reinforced nylon 66. In specimens exhibiting short lives, the matrix appeared
to fail by fatigue crack growth (radially about the fibers) and in a relatively
brittle manner (Fig. 5.32). In contrast, with specimens exhibiting long lives,
failure was clearly more ductile (Fig. 5.33); at the same time, fiber–matrix
adhesion was evidently strong (see also Cessna *et al.* [99]). This finding
suggests that one must be cautious in generalizing about material fatigue
response in composites, for these differences must reflect differences in the
nature and distribution of flaws that are introduced adventitiously during the
fabrication of nominally similar composites.

Statistical Aspects of Fatigue. A quite comprehensive study of fatigue
response in a short-fiber-reinforced composite has been described by Di-
Benedetto *et al.* [112–115], who used nylon 66 reinforced with short carbon fibers and tested at 10 Hz in the tension–tension mode. Experimental
variables included: stress amplitude, mean stress, fabrication technique
(injection and compression molding, and annealing), and moisture con-

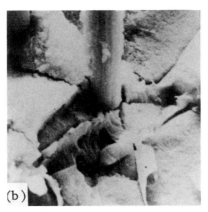

Fig. 5.32   Fatigue failure surface for carbon-fiber/nylon 66 composite with short fatigue life. (a) Magnification 190 × . (b) Magnification 2700 × . Note the low degree of plastic deformation and the indication of porosity. (From DiBenedetto *et al.* (112).]

tent (from 2.8 to 8.3% water); surface temperatures were also monitored. Specimen descriptions and loading conditions are given for one series of experiments in Tables 5.4 and 5.5. The overall purpose was twofold: to develop a statistical model for characterizing the reliability of the short-fiber composites, and to study the fatigue behavior (the latter topic will be discussed later). Related models have also been used with continuous fiber systems, see Section 5.3.2.

In the first study, a model was developed based on an approach proposed by Halpin *et al.* [116], which involved combining a Weibull distribution function with a power-law flaw growth model.

In the second study (Tables 5.4 and 5.5), DiBenedetto and Salee [114,

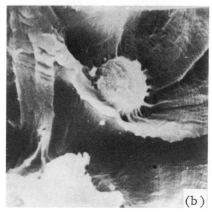

Fig. 5.33 Fatigue failure surfaces for carbon-fiber/nylon 66 composite with long fatigue life: (a) high degree of yielding (magnification 1000 ×); (b) cavitation induced by good adhesion (magnification 2700 ×). Note the radial propagation of a crack from the broken fiber. [From DiBenedetto *et al.* (112).]

TABLE 5.4 Experimental Populations of Carbon-Fiber Reinforced Nylon 66[a]

Designation	Treatment
$1A\frac{1}{2}$ RH 55	Nylon 66, fiber batch 1, compression-molded, annealed $\frac{1}{2}$ hr, conditioned at 55% RH
1A1 RH 100	Nylon 66, fiber batch 1, compression-molded, annealed 1 hr, conditioned at 100% RH
2A24 RH 55	Nylon 66, fiber batch 2, compression-molded, annealed 24 hr, conditioned at 55% RH

[a] From DiBenedetto and Salee [114].

TABLE 5.5 Stress Amplitudes in Fatigue Experiments with
Carbon Fiber/Nylon 66 Composites[a]

	Test Conditions			
Population	σ^{max} psi	σ_{mean} psi	$R = \sigma_{min}/\sigma_{max}$	$\sigma_{max}/\bar{\sigma}_B{}^b$
1 A 1 RH 100-1	9061	5458	0.20	0.83
1 A 1 RH 100-2	9279	5458	0.18	0.85
1 A 1 RH 100-3	9824	5458	0.11	0.90
1 A $\frac{1}{2}$ RH 55	11250	6250	0.11	0.90
2 A 24 RH 55	13566	8172	0.20	0.83

[a] From DiBenedetto and Salee [113].
[b] $\bar{\sigma}_B$ is the average stress-to-break.

115] found that the cumulative distribution of tensile strengths followed
a simple Weibull function:

$$P(\sigma_B) = \exp -(\sigma_B/\hat{\sigma}), \qquad (5.1)$$

where σ_B is the breaking stress, $P(\sigma_B)$ the probability that a specimen will
survive a stress σ, and $\hat{\sigma}$ a scaling factor that characterizes the average
strength. The distribution of fatigue life could also be described by a Weibull
function that was modified to account for the fact that some specimens
broke on the first half-cycle (σ_{max} being close to the average strength—see
Table 5.5) and that some survived 10^6 cycles. Thus the following function
held for the probability of time-to-break, $P(t_B)$:

$$P(t_B) = [1 - G(0)] \exp -(t_B/\hat{t})^{\alpha_f} + P(\infty), \qquad (5.2)$$

where $G(0)$ represents the fraction broken on the first half-cycle, $P(\infty)$
represents the fraction of specimens surviving beyond 10^6 cycles, $\hat{t}$ is a
scaling factor, and α_f is a constant. (The wearout model of Halpin et al.
[116] represented data well, but was sensitive to environment, and gave
physically unrealistic parameters in some cases.) As shown by Fig. 5.34,
Eq. (5.2) represents the data quite well (>95% confidence level). Since the
distributions of both tensile strength and fatigue life follow Weibull-type
functions, it is not surprising that strength and fatigue life can be correlated (see Fig. 5.35). (See also the correlation between static and fatigue
toughness discussed in Section 3.8.7.) By calculating corresponding values
of σ_B and t_B at 10% intervals along the distribution curves, beginning at
$P = 0.1$, the following master curve was obtained:

$$\log_{10}\sigma_B = A\left[\frac{\sigma_{max}}{\langle\bar{\sigma}_B\rangle}(1 - R)^{0.5}\right]^4 \log_{10} t_B + B, \qquad (5.3)$$

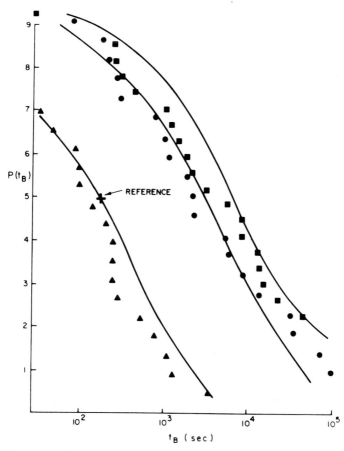

Fig. 5.34 Cumulative distribution of fatigue life in carbon-fiber/nylon 66 composites. Solid points represent experimental data: ■, 1A1RH100-1; ●, 1A1RH100-2; ▲, 1A1RH100-3; curves represent Eq. (5.2). [From DiBenedetto and Salee (114).]

where R is the ratio of minimum to maximum stress, A and B are constants (depending on the fiber batch), $\bar{\sigma}_B$ is the *average* strength of the population and the other symbols have the same significance as before. While the slopes of the master curves (which correspond to the degree of dispersion of the distributions) are similar, the intercepts are not. This difference in scaling presumably reflects a difference in the fiber strengths of the two batches. In any case, the correlation of strength with fatigue life should provide a useful way of predicting fatigue life as a function of service conditions that would affect strength, e.g., in this case, moisture content.

Not surprisingly, the behavior of injection-molded specimens was different; these specimens contained a higher proportion of fibers (14%

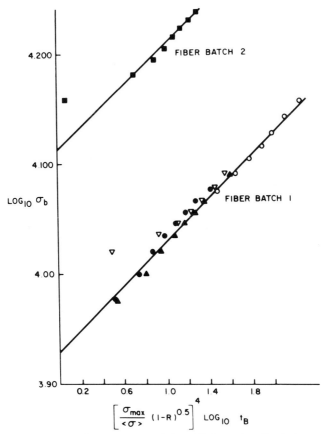

Fig. 5.35 Master curves of strength versus fatigue life for carbon-fiber/nylon 66 composites expressed in terms of Eq. (5.3), with $\langle\sigma\rangle \equiv \bar{\sigma}_B$. □, 2A24RH55; ○, 1A$\frac{1}{2}$RH55; ▲, 1A1RH100-2; ▽, 1A1RH100-3; ●, 1A1RH100-1 [From DiBenedetto and Salee (114).]

instead of 10%), the fibers were preferentially oriented, and the composites were more ductile. A Weibull-type function worked well for the distribution of breaking times and for the correlation of fatigue life with strength, but Eqs. (5.2) and (5.3) were modified [115] to include a third parameter, an induction time for the onset of fracture. The distribution of fatigue life for the compression-molded specimens was also quite different from that found for the injection-molded specimens. Compression-molded specimens showed generally better properties at longer times, but a tendency toward failure of a greater fraction at short times. Thus, *reliability* should be considered as a characteristic quite different from *average lifetime*.

Failure modes differed greatly, depending on the method of fabrication. The injection-molded specimens exhibited nonlinear and ductile behavior, the ductility being enhanced by hysteretic heating. In contrast, the compression-molded materials were linear in stress–strain response and failed in a brittle manner characteristic of fatigue crack propagation. These observations are consistent with the fact that fibers in injection-molded specimens exhibit preferential orientation, in this case with respect to the stress axis.

Fatigue Crack Propagation in Discontinuous Fiber Composites. Interestingly, in spite of the discontinuities introduced by the presence of short fibers, fatigue crack propagation in discontinuous-fiber systems generally follows the Paris crack propagation law (i.e., $da/dN = A\Delta K^m$). Thus Thornton [117] found linear relationships between $\log da/dN$ and $\log \Delta K$ (Fig. 5.36) for single-edge-notched specimens (tension–tension loading)

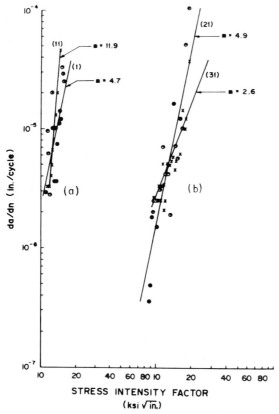

Fig. 5.36 Log growth rate (da/dN) versus log stress intensity factor range (ΔK) for discontinuous aluminum-fiber/epoxy composites at two values of load. Conversion: 1 ksi in$^{1/2}$ = 1.1 MPa·m$^{1/2}$. [*J. Compos. Mater.* **6**, 147 (1972), P. A. Thornton. Courtesy of Technomic Publishing Co., Inc. 265 Post Road West, Westport, CT 06880.]

of a polyamide-cured epoxy containing 13% (by volume) of 0.1-mm-diameter aluminum fibers (nominally randomly dispersed). At the same time, considerable variability in slope was noted. By scanning electron microscopy, it was found that high values of slope were associated with the occurrence of air bubbles, poor wetting of the fiber by the matrix, and a tendency toward bundling of the fibers. In contrast, low values of the slope were associated with good interfacial wetting, a high degree of randomness of fiber orientation, and relative freedom from air voids.

While the detailed response thus depended strongly on the quality of fabrication, two important observations were made. First, incorporation of the fibers resulted in a significant change in fracture behavior; the unmodified resin failed catastrophically, with little evidence of stable crack growth, at loads as low as 100 lb (45.4 kg). Also, the FCP curves with a mean load of 300 lb (136 kg) resembled those of an ABS resin (see Section 3.1); similar results were obtained with a mean load of 400 lb (180 kg).

The use of fracture mechanics techniques (see Chapter 3) to characterize FCP in short-fiber composites has also been demonstrated by Owen and Rose [102] and by Owen and Bishop [118] in studies of a chopped-strand-glass-mat/polyester (C) and a glass–cloth/polyester laminate (A). (Previous studies on the use of fracture mechanics to characterize K_c are reviewed by Wu [92] and by Holdsworth *et al.* [119].) Since with a composite, the "crack" is not a simple entity, but rather a summation of flaws that cannot be readily characterized in terms of length [71, 119], Owen and Bishop used a compliance technique to determine an *equivalent* crack length. In this way, difficulties with the visual observation of cracks were avoided. As reported earlier [103, 117], the Paris law was followed. From the FCP data in Fig. 5.37, it is clear that material C (slope $m \sim 6$, a value comparable to that of some polymers) is more resistant to FCP than material A ($m \sim 13$, an exceptionally high value for polymers). This difference must reflect the fact that in C a higher proportion of fiber bundles tends to lie transverse to the crack than is the case with the cloth-reinforced material. (For predictions of fatigue life from FCP data for these materials, see the following section.)

More recently, studies of FCP in short-fiber-reinforced systems have been conducted by DiBenedetto *et al.* [113] and by DiBenedetto and Salee [114, 116], using nylon 66 matrixes and carbon and polyaramid* fibers (Fig. 5.38). In the earliest study (on carbon-fiber-reinforced nylon 66), the Paris relationship was found to hold with $da/dN \propto \Delta K^{12}$ [113]. This exceptionally high power may be compared with the case of epoxy resins ($n = 10$–20) (see

---

* Aromatic polyamides, e.g., Kevlar.

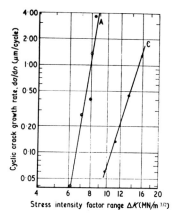

Fig. 5.37 Cyclic crack-growth rate (da/dN) plotted versus the stress intensity factor (ΔK) for glass-cloth/polyester laminate (A) and chopped-strand-glass-mat/polyester laminate (C) [Owen and Bishop (118), copyright by the Institute of Physics].

Ref. [120] and Section 3.8.3), Al-fiber-reinforced epoxy ($n = 3$–12) [117] and glass-fiber-reinforced polyester [119]; in all cases, n is higher than is usually observed for typical unmodified polymers (see Section 3.). In the subsequent study using similar carbon–nylon composites (Table 5.4), the effects of mean stress and the existence of lower and upper thresholds were first taken into account by combining a model proposed by Erdogan [122] for fatigue crack propagation was combined with a model proposed by Kitagawa and Motomura [123] for crack growth in viscoelastic solids. However, the resulting five-parameter equation was found to be no better than the following simpler equation (Fig. 5.38, Table 5.6):

$$\dot{a} = b[K_{max}(1 - R)^c]^r = b[K_{max}^{1-c}\Delta K^c]^r, \qquad (5.4)$$

where $\dot{a}$ is the rate of crack growth per unit of time, b, c, and r are constants, R is the ratio of minimum to maximum stress, and $K_{max} \equiv \Delta K_{max}$ if $K_{min} \sim 0$. Results showed that c and r varied between 0.1 and 0.4 (c) and 16 and 19 (r). FCP rates were enhanced by the presence of 8% water (compared to 2.8%), in this respect resembling the case of the same nylon matrix discussed above (see Section 3.8.4). FCP resistance was also greater, the higher the strength of the fiber.

As expected, fractographic examination revealed that fibers oriented perpendicular to the stress direction were the first sites of damage (cf. the case of other examples cited in earlier discussions, e.g., Owen and Dukes [107] and DiBenedetto *et al.* [112]). However, in the case of strong interfacial adhesion, fiber debonding was not dominant; instead, the matrix appeared to

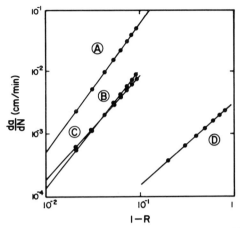

Fig. 5.38 Crack propagation velocity (at $\Delta K_{max} = 6.32$ MPa·m$^{1/2}$) in discontinuous-glass-fiber/nylon 66 versus $1 - R$, where R is the ratio of minimum to maximum load [DiBenedetto and Salee (113)]. For descriptions see Table 5.6.

undergo radial drawing about the fiber, in a manner similar to that observed in crazing. In addition, fiber pullout appeared to be important.

Although tests were not performed on the unmodified matrix, it is interesting to compare results for the composites with the behavior observed by Bretz *et al.* [121] for the same nylon 66. With the random-in-plane fibers, crack growth rates listed in Ref. [113] were about an order of magnitude lower for a water content of 2.8% than for a water content of $\sim 8\%$, in general agreement with the findings of Bretz *et al.* [121] that small amounts of water increase the FCP resistance of nylon 66.

The effect of short carbon fibers on FCP (at 6 Hz) of nylon was also compared with the effects of polyaramid fibers and hybrid carbon/polyaramid systems, the total fiber concentration being 20 vol% in each case [113]. (Unfortunately, the authors do not state which of the two carbon-fiber batches referred to in Table 5.4 was used in this study.) Specimens were fabricated with polyaramid and carbon fibers to yield random-in-plane fiber configurations; in addition, a hybrid laminate was prepared by alternating layers of random-in-plane graphite and polyaramid composites (see Table 5.6). (A random-in-plane hybrid was also prepared and tested, but data were not compared with those for the other composites.) As in the previous study, water contents of 2.8 and 8.3% were obtained by appropriate conditioning of the specimens.

Although there are insufficient data to permit full elucidation of the effects of fiber content and frequency, several conclusions may be drawn (see Table 5.6):

TABLE 5.6 Fatigue and Stress–Strain Characteristics of Short-Fiber-Reinforced Nylon 66[a]

Code	Type (vol %)	Geometry	(water %)[e]	Tensile strength (MPa)[e]	E (GPa)	ε_B	b[f]	c[f]	r[f]	a (cm/min)[g]
A	Carbon (20)	Random-in-plane	2.8	127	8.5	0.02	1×10^{-51}	0.12	17.9	9.8×10^{-3}
B	Polyaramid[b] (20)	Random-in-plane	2.8	127	8.5	0.022	4.2×10^{-9}	0.79	2.4	6.6×10^{-3}
C	Carbon[c] (10) + polyaramid (10)	Laminate	2.8	136	8.8	0.017	1.3×10^{-20}	0.25	7.5	4.9×10^{-3}
D	Carbon[d] (10) + polyaramid (10)	Laminate	8.3	105	6.3	0.021	8.2×10^{-13}	0.47	2.9	2.7×10^{-5}

[a] DiBenedetto and Salee [113].
[b] Kevlar-49, type III.
[c] Each plane contained a random mixture.
[d] Alternating layers of random-in-plane carbon and polyaramid composites.
[e] 1 GPa = 145 ksi; 1 MPa = 145 psi.
[f] Constants from Eq. (5.4): $\dot{a} = b[K_{max}^{(1-c)} \Delta K^c]^r$.
[g] At $K_{max} = 600$ kg cm$^{-3/2}$ (6 MPa·m$^{1/2}$), $R = 0.5$.

(1) With random-in-plane fibers, the resistance to FCP is significantly higher for the polyaramid system than for the carbon one.

(2) At equal fiber concentration, and hence at approximately equal modulus, the hybrid composite was about equal in FCP resistance to the polyaramid system, and hence superior to the carbon system.

(3) The FCP resistance of the hybrid composite was higher when saturated with water (8.3%) than when equilibrated to 2.8% water content. (This result does not agree with the findings discussed earlier for carbon composites or with observations by Bretz *et al.* for nylon 66 itself [121].

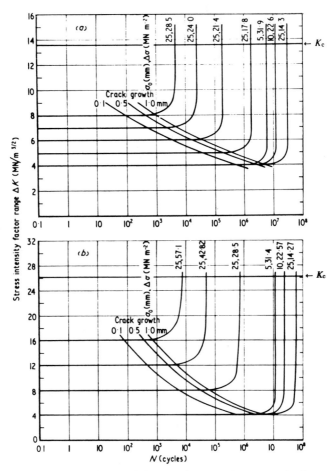

Fig. 5.39 Fatigue failure of glass-reinforced polyester under constant load range: (a) material A, $K_c = 13.6$ MPa·m$^{1/2}$; (b) material C, $K_c = 26.2$ MPa·m$^{1/2}$. Materials: (a) Chopped-strand-glass-mat/polyester, and (b) glass-cloth/polyester [Owen and Bishop (118), copyright by the Institute of Physics].

(4)   The sensitivity of FCP rate to the mean stress (as measured by γ) tended to be greater, the greater Young's modulus.

Fractographic studies of specimens broken in simple tension and in fatigue were used to estimate the ratio of fiber-pullout length to diameter, and hence the interfacial shear strength. Values of 1.8×10^4 psi (120 MPa) and 6.7×10^4 psi (460 MPa) were found for the carbon and polyaramid fiber in tension, respectively; values for specimens fractured in fatigue, though not calculated, would be higher since the pullout lengths were lower than those observed in tension. Significant yielding was noted in the nylon matrix as well as adhesion to the fibers, especially in the case of carbon fibers. With the polyaramid system, fatigue damage occurred first in the form of debonding at fibers that lay perpendicular to the stress axis. Debonding then continued, affecting fibers at progressively smaller angles to the stress axis, until eventually cracks appeared in the nylon itself. With the carbon system, a similar damage mechanism was suggested, with the added phenomenon of resin drawing along the stress axis, due to the high level of interfacial adhesion.

Prediction of Fatigue Life.   The use of S–N curves in predicting fatigue life has already been discussed. Owen and Bishop [118] also considered the prediction of fatigue life in such systems in terms of ΔK rather than stress. It was concluded that, at least with these systems, it should be safe to use data obtained from tests up to 10^6 cycles to predict behavior at 10^{10} or 10^{11} cycles, a prediction not possible with conventional S–N curves. For failure under a constant load range, the Paris expression was integrated and plots obtained of ΔK versus N (Fig. 5.39). Thus the technique described in Section 3.9 can be applied as well to discontinuous-fiber-reinforced plastics.

5.3.2   Continuous-Fiber Composites

While the incorporation of discontinuous fibers into a polymeric matrix can confer significant improvements in static and fatigue strengths as well as stiffness, the highest level of properties is obtained by the use of continuous fibers [124]. Hence continuous fiber systems are used for the most demanding structural applications; in other cases, discontinuous fibers or combinations of discontinuous with continuous fibers (as in sheet molding compounds) may be adequate. As with discontinuous fibers, many geometrical variations are used, including unidirectional and cross-plied arrays, woven cloth, and mats that have various fiber orientations. Fabrication is usually effected by such processes as laminating, filament winding, molding (for example using notched dies, vacuum bags, and hand layup), and pultrusion (extrusion into shapes by pulling fibers first through a resin bath and then through a die) [124, 125]. Although both thermoplastic and thermosetting (cross-linked) resins are used, thermosetting resins are used for composites subjected to the most severe service conditions because of their higher modulus and dimensional stability, especially at elevated temperatures. It should be noted that these properties of thermosetting resins are coupled with a generally lower fracture toughness and fatigue resistance than is generally the case with thermoplastics [120].

Of course, the principles and failure modes mentioned in Section 5.3.1 and in the Appendix apply to continuous fiber systems as well as to discontinuous ones. For example, hysteretic heating and dominance of the matrix with certain loading conditions must be taken into account. With laminates, additional complications arise because of restraints induced by the plies, and complex interactions between the stress field and the damaged zone [71]. In this section, typical behavior in terms of damage, fatigue strength, and crack or damage propagation is emphasized; the effects of temperature, frequency, and environment are also discussed. Because of space limitation and the fact that many questions remain unanswered, no attempt is made to review critically the entire voluminous literature or to provide a complete bibliography. However, a large-enough number and variety of references is provided to lead the reader to sources of more specific information and to additional examples (see, especially, [70, 75–83]).

General Aspects of Fatigue in Continuous-Fiber Composites. In both glass-fiber type and graphite-fiber-reinforced plastics (gfrp and cfrp, respectively), localized damage occurs at low stresses very early in the fatigue life (the term carbon is used interchangeably with the term graphite). McGarry [126] and Broutman and Sahu [127] noted a disproportionate amount of damage (in the form of fiber breakage) in a gfrp during the first few cycles; Fuwa *et al.* [128] made a similar observation for cfrp. As discussed below, such damage may or may not be reflected in major changes in performance, depending on the extent and on the system. To be sure, the significance of such early damage depends upon the extent it is reflected in changes in mechanical properties. In any case, complete understanding of the fatigue processes requires the characterization of damage throughout the fatigue life, and the relating of the damage to critical properties.

Thus many different methods have been used to characterize the damage in composites developed during cycling (for a review see Nevadunsky *et al.* [129]). While visual (including microscopic) examination is useful in deducing mechanisms of failure, much attention has been given to providing more quantitative measures of damage. Of these the most operational include the characterization of decreases in strength or modulus [128]. However, since strength tests are destructive, much interest in nondestructive tests has developed. For example, strain gauges have been used to follow the development of off-axis changes in compliance during tests [129]; in some cases, changes in damping (as in Fig. 5.40) have been used to monitor the progress of damage [129–133]. With many composites, x-ray techniques (in which, typically, a radio-opaque additive such as tetrabromoethane is injected at a crack) have proved to be useful in revealing details of damage at flaws [134–136]; perhaps most attention has been given to the use of a variety of ultrasonic techniques [129], which are sensitive to fairly large voids, cracks, and debonded regions. Although there are often problems with sensitivity and with interpretation, recently developed ultrasonic techniques show promise for improvement. Approaches based on the determination of dispersion and velocities of ultrasonic waves and pulses are receiving attention [137], and the use of a diffraction model to correlate damage with attenuation in pulse-echo determinations has been proposed [137] (Fig. 5.41). Acoustic emission techniques are also being used to an increasing extent. (For a review see Williams and Lee [138]; as fibers break, acoustic emissions [pulses of noise] occur [Fig. 5.42].) In addition to the emissions count, the amplitude distribution and frequency spectrum may provide useful information [139]. With a well-made cfrp composite, emissions occur on the first cycle, and as the peak load is reached during subsequent cycles [128], however the number of emissions falls progressively until catastrophic failure

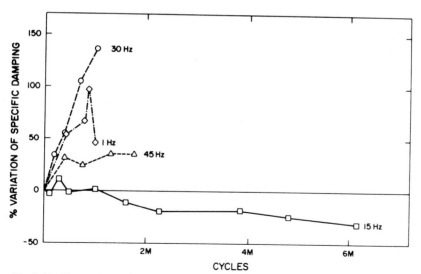

Fig. 5.40   Change in specific damping of a boron/epoxy laminate at four test frequencies. Note the complex effect of frequency. [Stinchcomb *et al.* (132). Reprinted with permission from the American Society for Testing and Materials, 1916 Race Street, Philadelphia, PA 19103 ©.]

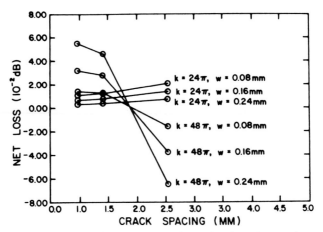

Fig. 5.41 Predicted changes of attenuation in ultrasonic pulse-echo experiments as function of crack width in graphite/epoxy laminates. Values are based on calculations using a diffraction model; trends are consistent with observations, though experimental values are lower than predicted. [Hayford and Henneke (137). Reprinted with permission from the American Society for Testing and Materials, 1916 Race Street, Philadelphia, PA 19103 ©.]

occurs (indicating that degradation of the matrix, fiber, or interface plays a major role throughout much of the fatigue life). With a less-sound cfrp composite, the rate of emission may not fall, and early failure occurs. In some cases, acoustic emission data have been correlated with macroscopic damage [138]. Whatever the method used, caution should be used in interpreting the significance of damage noted by sensitive nondestructive tests. The question is not simply whether or not damage has occurred, but how significant the damage is with respect to performance.

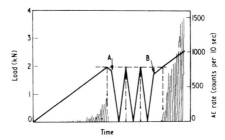

Fig. 5.42 Acoustic emission (AE) during several load cycles prior to loading to failure of cfrp prepreg. Note that emission begins just prior to the previous maximum load level and continues for a short time after unloading has begun. Gain, 80 dB; cross-head speed 0.05 mm/min except between A and B where it is 0.5 mm/min. [After Fuwa *et al.* (128), copyright by the Institute of Physics.]

Fatigue Life: S–N Curves. While the value of S–N testing in composites that do not exhibit a clear endurance limit may be questioned [70], nevertheless most fatigue testing has involved this kind of characterization. In any case, the residual strength revealed by an S–N curve can reveal the sensitivity of fatigue response to the type of loading, material composition and geometry, and the environment. Thus, unidirectionally oriented carbon and boron fibrous composites (cfrp and bfrp) cycled in axial tension give very flat S–N curves at high stresses [69] (see, for example, curve B in Fig. 5.43). Indeed, cfrp is better than aluminum in this respect [69, p. 352]. Below the scatterband (not indicated) fatigue failure is unlikely, though occasional failure due to time-dependent fracture may occur [128]. Figure 5.43 (curves B and C, and additional figures given below) also shows that when the load is not tensile, uniaxial, and aligned with the fiber axis, fatigue failure occurs at lower stresses; thus flexural, compressive, and shear stresses place severe loads on the matrix (and its interface). Although such stresses have not been used in fatigue studies as extensively as tensile stresses, they are to be expected in many loading conditions; if present, they tend to lower fatigue resistance by inducing buckling, debonding, delamination, and matrix cracking, the precise combination depending on the orientation of the fiber axes with the stress axis [87, Appendix]. The nature of the fiber is important too, carbon and boron fibers being more effective than glass in conferring fatigue resistance. This superiority may be due in part to higher thermal conductivities

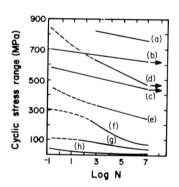

Fig. 5.43 An approximate comparison of the fatigue behavior of several reinforced plastics; values at the extreme left represent static strength data [After Harris (70), who also gives original literature citations for each curve.] (a) boron/epoxy laminate, 10 ply $0 \pm 45°$, axial tension cycling; (b) carbon (type 1)/polyester, unidirectional, $V_f = 0.40$, axial tension cycling; (c) as (b), repeated flexure cycling; (d) carbon/epoxy laminate, 18-ply $0 \pm 30°$, axial tension cycling; (e) as (d), axial compression cycling; (f) glass/polyester, high strength laminate from warp cloth, axial tension/compression cycling; (g) glass/polyester composite, laminate, tension/compression cycling; (h) polyester short-fiber molding compound, $V_f \simeq 0.12$, tension/compression cycling.

(which will tend to reduce hysteretic heating), and in part due to the lower matrix *strains* at a given applied stress, the latter effect being a consequence of the exceptionally high Young's moduli of cfrp, up to about 75×10^6 psi (500 GPa) (see Section 5.3.1).

Much of the early work on fatigue was performed by British groups, and has been reviewed by Owen [69]. Most of the tests were run in tension using uniaxially oriented fibers; the results of studies by, for example, Owen and Morris [140] and Beaumont and Harris [141] were the basis for curves such as B in Fig. 5.43. The excellent properties of cfrp systems compared to those of other materials were also demonstrated when stress amplitude was plotted against mean stress (at constant life, 10^7 cycles), both on an absolute and specific basis (Figs. 5.44 and 5.45). (This plot is often referred to as a Goodman diagram.) Curiously, the appearance of fracture surfaces of specimens broken in tension under static and fatigue loading were similar, possibly because, since high stress levels are involved, only a few fibers need fail before catastrophic fracture ensues. In compression cycles, failure was accompanied by longitudinal splitting and fiber buckling.

A detailed study of axial fatigue failure in unidirectional glass fiber composites (E-glass/polyester) has been recently described by Kim and Ebert [139] with emphasis on determining the sequence of failure events underlying the S–N curve. Damage was characterized not only by changes in modulus and microstructure, but also by changes in damping. Tests were run in tension, under load control, and at 10 Hz. It was possible to characterize

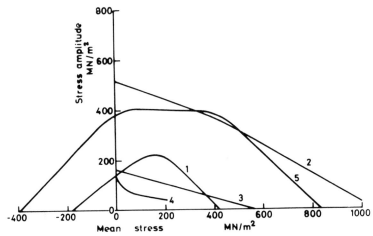

Fig. 5.44 Comparison of Goodman diagrams for various materials: 1. 0–90° cross-plied cfrp, 2. AISI 4340 alloy steel, 3. 7075-T8 aluminum alloy, 4. 0–90° cross-plied, glass-reinforced plastic, 5. unidirectional cfrp. [Owen and Morris (69, 140) who also cite the original sources of data.]

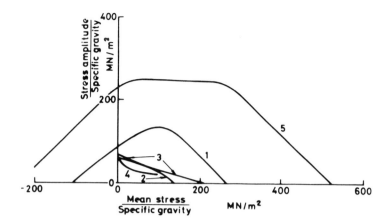

Fig. 5.45 Comparison of Goodman diagrams on a specific stress basis; data of Fig. 5.44 divided by specific gravity [Owen and Morris (69, 140)].

the failure microstructurally in terms of three stages (I, II, and III) and to correlate these stages with changes in modulus and hysteresis energy (cf. Fig. 5.46). (Note that the abscissa is the *fraction* of the fatigue life in cycles.) Significant effects of surface characteristics (as-molded versus machined) were also observed (see also Fesko [142]). Stage I (see Fig. 5.46) was characterized by a drop in modulus essentially to a steady-state stage II; this was associated with the propagation of surface flaws on the fibers (following significant damage on the first half-cycle). The drop in modulus, which was also reflected in "noise" in plots of damping versus time, was greater with machined surfaces, presumably because machining increases the population of surface flaws. Stage III corresponds to the final stages of cyclic life where the modulus ratio rapidly deteriorates.

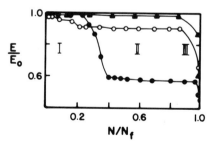

Fig. 5.46 Static modulus degradation behavior during fatigue tests for three different surface conditions and three failure steps (E/E_0, ratio of instantaneous to initial secant modulus; N/N_f, ratio of interrupted cycles to number of cycles to failure). [*J. Compos. Mater.* **10**, 139 (1978), H. C. Kim and L. J. Ebert, Courtesy of Technomic Publishing Co., Inc., 265 Post Road West, Westport, CT 06880.]

Kim and Ebert [139] then related these stages to the micromechanisms of failure by examining the fractured specimens. During stage I, the fiber surface flaws apparently coalesce and lead to interfacial failure and the onset of transverse and shear cracking in the matrix. During stage II, the static modulus remains nearly steady, but the hysteresis energy rapidly increases. Finally (stage III), the matrix cracks coalesce and fibers on a plane fail, causing delamination through the specimen cross-section, in a manner consistent with Table 1 (see the Appendix).

At least up to the early 1970s, most research was conducted in axial tension; however, some was done with more severe stress systems, such as in modes of loading more damaging than tension (e.g., compression, flexure, torsion, and interlaminar shear). Since, as mentioned above, such severe modes of loading may be expected to occur in real composites subjected to realistic service conditions, it is important to consider their effects. In

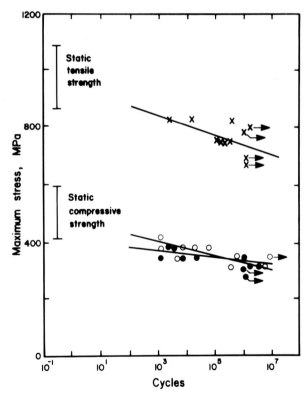

Fig. 5.47 Axial load fatigue results for unidirectional, surface-treated, high-modulus fiber composites prepared from prepreg tows with a typical epoxy resin system; volume fraction of fiber 0.61; 117 Hz; compressive stress plotted positive (lower curve). [After Morris (143).]

fact, the number of studies of these more severe modes has continued to increase, as realization of the importance of nontensile response has grown.

As expected, axial cycling that involves compression leads to a deterioration of fatigue life (Figs. 5.44 and 5.47); in the case of the cfrp studied by Morris [143], a given fatigue life could be obtained in compression cycling (at zero mean stress) only by keeping the stress one-half that of the stress corresponding to tension cycling. A combination of fiber buckling with debonding, matrix shear failure, and longitudinal splitting is to be expected in such cases (see the Appendix).

Static and fatigue strengths of unnotched cfrp laminates have also been studied by Schütz and Gerharz [144] at 20 Hz, over a range of stress ratios, and in compression as well as tension. Tests were run at both constant and variable load amplitudes, the latter to simulate a flight-by-flight load sequence typical for the wing-root area of an aircraft. The authors concluded that while cfrp composites are generally superior to, e.g., aluminum alloys, account must be taken of specific differences between metals and cfrp, especially the sensitivity of the latter to low cycle fatigue in compression. The authors also discuss the difficulties of developing fatigue stress design data for cfrp systems, e.g., the effect of the particular layup used.

Several studies of flexural fatigue have been made, for example, by Dharan [145, 146] and Fesko [142]. In an investigation of unidirectionally oriented carbon-fiber-reinforced polyester (fabricated by pultrusion),* Dharan [145] varied frequencies between 2 and 4 Hz in order to maintain a constant strain rate. Tests were run between fixed strain limits, some in a completely reversed bending mode and some in zero to tension/compression (the latter ensuring that one surface experienced only tensile, and the other compressive, forces). As has been noted above for other systems, fully reversed loading proved to be the more severe of the two modes employed. Especially when the fatigue life data (failure defined as a 15% loss in modulus) were plotted in terms of strain amplitude, there was remarkably little effect of cycling (up to 10^7 cycles) below a critical strain of about 0.004. In general, results agreed with those of Fig. 5.47, though Dharan's curves were displaced toward higher stresses. Above this critical strain, a fiber buckling instability on the compression surface appeared to be involved in the initiation of a crack. This was usually followed by delamination under the initiation zone, formation and propagation of a transverse crack, and delamination (or mode II crack

---

* Much of the research and development of undirectional fiber polymer composites has been directed toward high-performance systems such as carbon/epoxy composites for aerospace applications. Nevertheless, there is still much interest in systems that exhibit lower levels of mechanical properties but that still are suitable for other weight-saving but less demanding applications at reasonably low cost (e.g., for applications in automobiles). In any case, the performance of carbon fiber/polyester composites is superior to that of glass-based composites.

propagation parallel to the fibers). In all cases, failure was always initiated
in the compression zone (Fig. 5.48).

Later results [146] confirmed the findings just mentioned. A criterion for
fatigue failure was also proposed based on the assumption that the effective
cyclic strain must exceed a critical value that depends on the matrix fatigue
behavior. For the case of axial loading parallel to the fiber direction, the
criterion was expressed as follows:

$$\sigma_{ec} = \varepsilon_{em} E_c \qquad \text{when } \sigma_{fc}/E_c > \varepsilon_{em}, \tag{5.5}$$

$$\sigma_{ec} \approx (\sigma_{fc})_{min} \qquad \text{when } \sigma_{fc}/E_c < \varepsilon_{em}, \tag{5.6}$$

where σ_{ec} is the composite long-life fatigue stress, ε_{em} the matrix long-life
fatigue strain, σ_{fc} the (minimum) composite static fracture stress, and E_c the
composite modulus. For angles other than zero between the applied stress
and the fiber direction, the matrix is subjected to shear strain in addition to
normal strains, and an equivalent strain should be calculated depending
upon the particular yield theory that applies. In any event, the fatigue life
will be enhanced if the applied cyclic strain in the matrix is kept below a
maximum value.

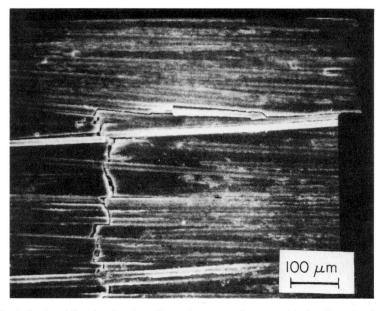

Fig. 5.48 A unidirectional carbon-fiber polyester specimen tested in bending, showing the
top (compression) surface in which both transverse and delamination cracks are present. Both
cracks go into the paper [Dharan (145), © 1974 AIME, New York].

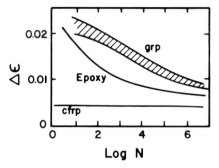

Fig. 5.49 Variation of cyclic strain amplitude during constant load cycling of gfrp, cfrp, and the epoxy resin on which both composites are based. The scatterband for gfrp contains data for composites containing fiber volume fractions from 0.16 to 0.50; the cfrp data are for composites with fiber volume fractions of 0.33 to 0.50. All composites are unidirectional, filament-wound material. [After Dharan (147) © 1976 AIME, New York.]

Indeed, plotting cyclic *strain* amplitude (from constant-load tests) against the number of cycles reveals clearly the consequences of carbon's exceptionally high modulus, which implies correspondingly low strains at a given load [147]. The data in Fig. 5.49 support the earlier idea [146] that fatigue failure requires that the maximum effective cyclic strain amplitude in a constant-load test* must exceed a critical value that depends on the matrix.

In another study of flexural fatigue, the behavior of pultruded glass/polyester composites was described by Fesko [142], who compared results with those obtained for several other unidirectionally oriented glass-fiber/polyester composites. Tests were constant amplitude, with failure defined as a 5% drop in peak load; both wet and dry environments were used. It was found that fatigue strength (at 300,000 cycles) was relatively constant (at 400 MPa) from 50 to 60 vol% of glass, and that fatigue stresses resembled those for compression-molded specimens. Moisture, while deleterious to fatigue strength, had less effect than with cross-plied composites. Machining affected behavior adversely due to the breakage of fibers thus induced (see also Kim and Ebert [134]). Since the specific loads used corresponded to relatively large strains ($\sim 1\%$), it was concluded that these materials should be capable of withstanding at least low-cycle fatigue applications.

Dharan [148] also studied fatigue life in glass-fiber/epoxy systems, and was able to divide the S–N curve into three regions, corresponding to dif-

---

* One must be careful to distinguish between the strain amplitude *corresponding* to the stress amplitude in a constant-*load* test and the strain amplitude *applied* in a constant-*strain* test. (See also the discussion in Chapter 2.) Unfortunately, the distinction is not always made clear, especially in abstracts of papers.

ferent types of damage. (It should be noted that this is a different division from that proposed by Kim and Ebert [139], whose stages corresponded to the latter part of the fatigue life (the last decade or so).) The three zones may be characterized as follows:

(1) At very early stages: relatively little effect of cycling on strength; failure based on fiber mean strength and distribution.

(2) Intermediate stage (200 to 10^6 stress reversals): log stress decreases with log number of cycles; failure by growth of matrix microcracks, fiber failure, and interfacial shear failure.

(3) Final stage: no failure, stress apparently below that required to propagate a crack through the fibers (in other words, the S–N curve became asymptotic).

A dominant role of the matrix in determining fatigue life is especially evident in shear loading, as shown by results of tests conducted under either interlaminar shear [149, 150] or torsional conditions [141, 149], the latter being more severe than the former. It has been deduced that failure in torsion involves debonding at the fiber–matrix interface, followed by the initiation and propagation of macroscopic cracks at some characteristic time [149].

So far the discussion has been concerned with undirectional fibers. In practice, many composites are based on woven cloth or laminates having plies at various angles to each other, for example, at 0, ± 45, and 90°. By appropriate choice of angles, one can minimize the tendency of a single type of crack to propagate catastrophically. Examples of detailed studies of fatigue life and related questions of various fibrous systems include: testing in zero-tension or tension–tension [150a, 151–155], compression or tension–compression [150a, 152, 156], and interlaminar or torsional shear [153, 157, 158]. Fibers studied include boron [154, 157, 158], carbon [152, 153, 155], glass [153, 156, 157], polyaramid [149, 159], and hybrids [151, 160]. While one may expect to see the general aspects of behavior typical of undirectional systems [for example, the effect of the angle between the fiber and stress axes (see Appendix)] there are, as mentioned above, additional complications due to the coupling of responses and to ply constraints [71]. Thus the interplay between stacking sequence and various failure modes may be very important, even though our analytical understanding is limited [158].

Statistical Aspects.   In general, the residual strength of fibrous composites drops during cycling [161], the rate of decrease depending on the system; some interesting exceptions are discussed later. (For a comprehensive review of fatigue life prediction, see Hahn [162].) Since the normal scatter in static strength data of fibrous composites is often increased by the random accumulation of damage (especially in glass-fiber systems), it is important to consider the statistical significance of residual strength [117, 161]. As mentioned above, Halpin *et al.* [116, 161] introduced the concept of wearout to characterize the failure of fibrous composites. This concept is based on the reasonable assumption that the accumulation of damage during cycling causes the decrease of strength until the residual strength and the applied stress are equal; as shown in Fig. 5.50, a significant statistical variation (probably greater with glass than with carbon fibers) exists. (Of course, this concept may be useful to particulate or discontinuous fiber systems as well, and to any other systems in which considerable random damage occurs during fatigue.)

While Halpin *et al.* [116] proposed an equation for the degradation of strength based on crack propagation model typical of homogeneous materials, it is now well recognized that the damaged zone is much more complex than a crack [71]. For this reason, Hahn and Kim [163] introduced

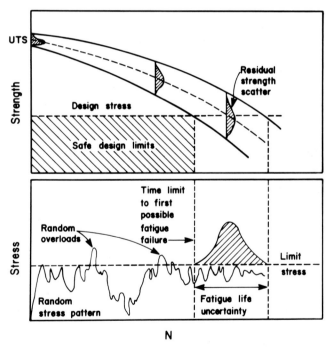

Fig. 5.50 Schema showing the design problems found with glass-fiber composites under fatigue loading. Note the interaction between reduction in tensile strength, increased variability in strength, and fatigue life uncertainty [Harris (70), after a diagram by Halpin *et al.* (161) Reprinted with permission from the American Society for Testing and Materials, 1916 Race St., Philadelphia, PA 19103 ©.]

the concept of characterizing the rate of change of residual strength, without reference to a single crack. Assuming that the rate of strength decrease over time was inversely proportional to some power of the residual strength, and taking into account the distribution of static strengths, they derived an expression for the fatigue life distribution. In their treatment, it was assumed that the rank of a specimen in fatigue strength will be the same as the rank in static strength. Making use of this implication (the "strength-life equal rank" assumption), Chou and Croman [164] have introduced a third parameter to allow for the existence of different distributions in residual strength:

$$F_{R(n\gamma_1)}(y) = 1 - \exp[-x_\gamma^\alpha(y) + x_{\gamma_1}^\alpha], \tag{5.7}$$

where $F_{R(n\gamma_1)}(y)$ [6] is the distribution of residual strength, x_γ the maximum static strength, y refers to the residual strength, and α a constant. Chou and Croman concluded that two processes are present during fatigue, one a tendency of the mean strength to decrease, and the other a tendency of residual mean strength to increase due to the weeding out of weak specimens. Indeed, populations were found in which one or the other of the two processes was dominant. Thus, in some cases, the mean *fatigue* strength may exceed the mean *static* strength.

Weibull distributions have also been frequently used. For example, Sims and Brogdon [153] and Awerbuch and Hahn [165] have reported that a two-parameter Weibull distribution was adequate to characterize the distributions of strength and life, for carbon-fiber/epoxy and glass-fiber/epoxy laminates and unidirectional carbon-fiber/epoxy composites, respectively. However, Ryder and Walker [152] found that a three-parameter distribution function was required to obtain the best fit for their data on carbon-fiber/epoxy laminates. A three-parameter function was also described by Yang and Liu [166], and a modified Weibull function has been proposed by Nakayasu [167]; for another treatment, see Harlow [167b].

Although the statistical treatments discussed so far have been concerned only with unnotched systems, considerable attention has been given to the use of notched laminates. McLaughlin *et al.* [168] developed a model to treat axial loads as well as colinear crack propagation for notched laminates. Another model has been proposed by Yang and Jones [169]; this model is equivalent mathematically to the Yang–Liu model for unnotched specimens [166]:

$$F_{R(n)}(x) = 1 \exp\left\{-\left[\frac{x^c + \beta^c KS^b n}{\beta^c}\right]^{\alpha/c}\right\}, \quad x \geq \sigma_{\max}, \tag{5.8}$$

where $F_{R(n)}(x)$ is the distribution function for the probability that the residual strength after n cycles $R(n)$ is smaller than the value x, α and β are shape and scale parameters, S is the stress range; and b, c, and K are constants. Results of axial shear tests on carbon-fiber/epoxy laminates gave good agreement with predicted values.

The model derived by Yang and Liu [166] was later extended to account for shear and compressive loads [169, 170] and various load sequences [171] and found to give predictions for unnotched specimens that agreed well with experiment.

At the beginning of this discussion, it was pointed out that residual strength does not *always* decrease on cycling. While strength is *usually* decreased by cycling, paradoxical improvements in strength may occur in some cases, especially in cfrp [92, 128, 139, 155, 165]. For example, Wu [92] has shown evidence for increases in fracture toughness in a unidirectional composite and has proposed that slightly misaligned fibers may span the growing crack and hinder its progress. Others have found evidence for strengthening due to development of a more diffuse damaged zone, which may, for example, involve the blunting of transverse cracks due to delamination along the load axis [158, 172, 173].

For a detailed discussion of recent statistical models for the failure of composites, see the recent review by Phoenix [170a], who also proposes a model of his own based on the concept of a chain of fiber bundles.

Proof Testing. An interesting application of statistical ideas is that when a specimen survives a high static load, it will survive some minimum number of cycles [163, 166]. In practice, this phenomenon is the basis for the *proof testing* of a component in order to guarantee a certain service life subsequent to survival of the test. Also, the service life can in some cases be extended by periodically applying a prescribed high load to composites during service [166]. At the same time, there is justifiable concern over the initiation of damage during the test itself [163]. It is true that there is some evidence that degradation of strength may be negligible if proof tests are conducted with loads $\leq 90\%$ of the mean ultimate strength [166], or, if damage is not visible, that the prediction may be possible with 90% confidence. However, even if crack blunting occurs during proof testing, crazes may be initiated that could make the composite more susceptible to environmental degradation. Thus much remains to be done to verify the validity of proof testing for various kinds of composites and flaws. For further details, see Hahn and Kim [163] and Yang and Liu [166], and, in the case of flawed structures, see Porter [174].

Fatigue Crack or Damage Propagation Rates. Several investigators have studied the propagation of damage from notches and holes in laminates; several examples may be noted. Kam [175] has studied the stresses required to initiate damage growth from bolt holes in cfrp composites. Sturgeon [155] observed that a central hole in a $0 \pm 45°$ cfrp reduced the monotonic strength to 76% of the integral material, but had little effect on fatigue life (in zero-tension tests) other than an increase in the variability of results. At the same time, the hole did initiate damage along the $\pm 45°$ plies, and the damage was reflected back along the balancing plies. The conclusion that notch sensitivity in fibrous composites was low agrees with an early report by Davis *et al.* [176] and, more recently, with the findings of Porter [174] for cfrp laminates of various kinds containing various kinds of flaws. On the other hand, Brinson *et al.* [177], who studied the growth of damage of notches and holes of various kinds in cfrp, found evidence for both stable crack growth and notch sensitivity. Owen and Bishop [178] also studied the ability of holes to initiate damage and found variable effects on residual strength, depending on the type of composite. Effects of impact loads on fatigue damage have also been examined [179, 180].

In view of the fact that a growing crack in a composite is more complex than in a homogeneous material, it may well be desirable to consider approaches other than that of Eq. (3.8). In other words, linear elastic fracture mechanics (LEFM) was not proposed for use in heterogeneous systems. While LEFM can often be used empirically (as above) even beyond its formal limits, other approaches may be advantageous in some cases. Thus Carswell [181] has proposed a model for the growth of a damaged zone from a stress raiser as follows:

$$dD/dN = C\sigma(l + pD)D/qN, \qquad (5.9)$$

where D is the dimension of the damaged region transverse to the stress direction, l the original length of the stress raiser, dD/dN the rate of damage accumulation per cycle, σ the cyclic stress range, and C, p, and q are constants. Redefinition of the crack as a process or damaged zone has also been proposed by Kanninen *et al* [182] in terms of the embedment of a local heterogeneous region into an anisotropic elastic continuum; a preliminary model was proposed to characterize fatigue crack growth (see also the discussion by Reifsnider [71]).

However, in spite of the complexity of the damaged zone ahead of a notch, it is apparently possible, at least in some cases, to predict fatigue lifetimes based on the concept of a more or less conventional crack. Thus earlier research on the failure mechanisms in laminates was extended by Mandell and Meier [183] to predict FCP rates in notched glass-fiber/epoxy laminates under tension–tension loading. Assuming (1) that the ligament at the crack tip is fatigued according to the S–N curve of the unnotched material, (2) a particular stress field, and (3) the validity of Miner's law, a model was developed for crack growth from an initial notch and shown to hold well for the cross-plied laminates studied. In this work, crack growth was found to follow a Paris relationship [Eq. (3.8)], with an exponent of 11. (See also the study by Owen and Bishop [118] discussed earlier; these authors applied a damage-zone factor to correct for the damage extending beyond the nominal crack.)

The growth of a notch was also studied by Campbell and Cherry [184], who characterized fatigue crack propagation rates in unidirectional glass-fiber/epoxy composites (Fig. 5.51) parallel to the fibers in the forward shear mode by an equation analogous to the Paris equation and having the form:

$$da/dN = (C/f)[G_{max}/G]^m, \tag{5.10}$$

where G_{max} is the maximum strain energy release rate, G the minimum strain energy release rate required for stable crack growth, m a constant (~ 5),

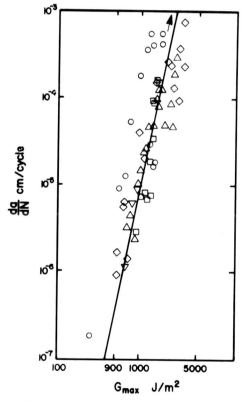

Fig. 5.51   Variation in fatigue crack propagation rate of unidirectional glass-fiber/epoxy composition with maximum strain energy release rate during the loading cycle. Test run in forward shear mode. [After Campbell and Cherry (184).]

and f the frequency. (For a more detailed discussion of frequency effects, see below, and also Section 3.4.) Equation (5.10) is, of course, essentially equivalent to Eq. (3.10), with allowance for an inverse dependence of FCP on frequency. By comparison with monotonic failures, it was suggested that the mechanisms were qualitatively similar, though quantitatively different due to greater hysteretic heating in the case of cyclic loading.

Fatigue crack propagation rates in the zero compressive mode were also studied by Kunz and Beaumont [185] with graphite-fiber-epoxy composites (both uniaxial and cross-plied), in air and in saline solution. Cracks grew initially, but then decelerated due to the occurrence of stress-relieving damage parallel to the axial fibers; after continued cycling, the crack speeded up again and broke through fibers in its path. The process of deceleration followed by reacceleration could then be repeated several times.

Effects of Temperature, Frequency, and Environment. Curiously, relatively little information is available on the effects of temperature on fatigue in polymers (see Section 3.5) and their composites. Of course, temperature may have several effects: on the viscoelastic state, on fracture toughness, and on chemical stability. In general, the effects of increasing ambient temperature appear to be detrimental; the higher the temperature, the greater the damage [185a]. This is not to say that the degradation is necessarily linear with respect to temperature; Worthington [186] has noted that the threshold fatigue stress level in hoop-wound cfrp rings tested in the tension–tension mode remained constant up to about 120°C, but decreased sharply thereafter. In an example of the effect of low temperatures, the tensile and comprehensive fatigue behavior (in the range from 10^2 to 10^6 cycles) of boron-fiber/epoxy laminates was found to be no worse (and perhaps slightly better) at -179°C than at room temperature [187]. The deleterious effect of elevated temperatures is also consistent with the degradation of properties when hysteretic heating is important.

Such effects of test temperature should be distinguished from effects due to *heat treatment*. In an interesting example of thermal history manipulation, Sun and Roderick [154] have shown that it is possible to introduce residual compressive stresses in boron-fiber/epoxy laminates by annealing at an elevated temperature under load, and then cooling. The resultant residual compressive stresses presumably reduced the matrix tensile stress under subsequent tensile loading, and thus significantly increased fatigue life, by at least an order of magnitude.

As with bulk polymers, frequency effects must be considered in composite systems. Thus with cross-plied boron–epoxy laminates tested under strain control, Stinchcomb *et al.* [132] have reported nonlinear effects of frequency on fatigue strength, stiffness, and damping, with maxima or minima at various frequencies; nonlinear effects have also been reported by Sun and Chan [137a] with notched graphite-epoxy laminates tested under load control (see Fig. 5.52). On the other hand, Reifsnider *et al.* [188] found that the damage (as reflected in modulus decreases) caused by cycling under load control was more severe, the lower the frequency. Inspection revealed that low-frequency cycling gave sharper, less diffuse damage zones than at higher frequencies.

Effects of frequency may also be related to differences in time-under-load (see Sendeckyj and Stalnaker [188a]). Similar adverse effects of low frequencies in composite systems and adhesive joints have been reported (for example, Campbell and Cherry [184] and Section 5.4) as well as in polymers themselves (Section 3.4). However, one might expect to find adverse effects at the high frequencies if hysteretic heating becomes sufficiently important to overcome the beneficial effects of time-under-load effect and damage-zone blunting. The nonlinear dependence of failure times reported by Sun and Chan [137a] may well reflect a balance of this kind. An adverse effect on going from 1 to 3 Hz was also noted by Rosenfeld and Gause [179] in a study of compressive fatigue in carbon-fiber/epoxy laminates using a standard load spectrum; the unfavorable effect was ascribed to a dependence

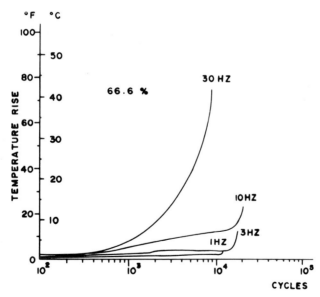

Fig. 5.52 Temperature rise for notched (center-hole) graphite-fiber/epoxy crossplied laminates under cyclic loads (66.6% of average static failure stress). The onset of rapid heating is correlated with failure [Sun and Chan (137a). Reprinted with permission from the American Society for Testing and Materials, 1916 Race St., Philadelphia, PA 19103 ©.]

on strain rate. Thus, as with polymers *per se*, effects of frequency are complex, and patterns of behavior not yet fully elucidated.

Earlier research by Romans *et al.* [189] on the effects of water on fatigue in composites was later extended by Gauchel *et al.* [156] to include large static exposure times (400 days) and to include copolymers of the brominated epoxy used by Romans *et al.* In fact, the fatigue life was inversely proportional to the initial water absorption of the matrix resin. Kunz and Beaumont [185] also reported complex effects of an aqueous saline environment on FCP in uniaxial and cross-plied graphite-fiber composites. Exposure enhanced axial cracking in all cases, and transverse ply splitting in cross-plied systems; at the same time, effective crack blunting and arrest occurred frequently due to interfacial failure. The deleterious effect of water may also be enhanced by high temperatures [190]; these authors also discuss in detail possible damage mechanisms.

At the same time, little specific effect of moisture on the behavior of graphite fiber and graphite/glass-hybrid composites has been reported by several investigators [141, 151]. However, these tests were conducted by preconditioning with moisture, and best results were obtained with 0° orientation of the fibers, less severe conditions than for total immersion or for cross-plied systems.

For more details on environmental effects, see also Bascom and Cottington [191], Mandell [192] and references cited in Section 3.4.

5.4 MISCELLANEOUS SYSTEMS

In addition to the composites discussed in previous sections, several additional systems that are not always classed or treated as composites deserve attention. For example, foil laminates are often excluded from consideration as composites. At the same time, polymer-bonded laminates do constitute a limiting case of a composite structure, and do exhibit the essential features of other composites considered previously: substrate, matrix, and interfacial region (the latter playing an increasing role, the thinner the bond). Indeed, since the problems and approaches to characterization of mechanical behavior (including fatigue) are essentially similar to those of other composite systems, it is useful to consider adhesive joints as composites, at least for the purpose of this monograph. For similar reasons, the relatively new classes of composites based on the incorporation of polymers into concretes of various kinds are also briefly considered. However, although polymeric coating/substrate systems could also be considered as constituting a limiting case of a laminate, they are not considered here, for they are not used as structural elements. At the same time, fatigue may well play a role in the degradation of coatings, and the principles used in the understanding of fatigue should be valid with coatings as well as with structural materials.

5.4.1 Adhesive Joints

As adhesive bonding is being used to increasing extents in structural and other applications, e.g., in high-performance aircraft [193], it is important to know the effects of stress conditions and environment on the lifetimes of adhesive joints (for a critical review, see Anderson *et al.* [194]). In service, aircraft not only sustain static loads but also several kinds of dynamic loads, for example, cyclic loads due to pressurization and depressurization (low frequency), gust loads (intermediate-to-high frequencies), and buffeting (high-to-very-high frequencies [195]). Problems are not confined to aircraft; fatigue failures severely limit the lifetime of the rubber-coated fabrics used for the skirts and seals that provide the suspension of air-cushioned vehicles [196]. Unfortunately, the prediction of service lives is not yet fully developed. Thus, the ranking of joint strengths based on monotonic loading tests may not necessarily serve even qualitatively as a predictor of life under typical service conditions [195]. Indeed, as has been noted generally with fatigue failure in polymers, cyclic loading is a more severe condition than static loading [195, 197].

In view of the considerable success of fracture mechanics approaches to determining the flaw tolerance in metal structures, increasing attention has been given to the possibility of application to adhesive joints. (An excellent summary has been given by Anderson *et al.* [194].) While relatively little has been published on the use of fracture mechanics to characterize flaw growth in adhesively bonded systems, interesting and promising results have been reported by several groups. It may be noted especially that, at least in some cases, sensitivity to *crack propagation* under static loading can be correlated directly with service experience, while other tests (in shear or peeling modes) are not useful predictors [198].

The first point of interest is what rate law is followed for flaw (or crack) propagation in adhesive joints. As part of what is probably the most extensive study of fatigue failure in such systems, Mostovoy and Ripling [197] showed that sigmoidal curves were obtained when $\log da/dN$ was plotted against $\log \Delta G$ (see Fig. 5.53), where ΔG is the range in strain energy release rate (see Section 3.3). (These studies were conducted using contoured-double-cantilever-beam-type specimens.) At low values of ΔG, a threshold value for crack propagation was observed (ΔG_{th}, below which crack growth rates become vanishingly small) in runout tests (10^6–10^7 cycles). Above this

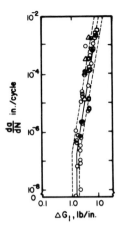

Fig. 5.53 da/dN versus ΔG_I for MRL-3 adhesive at room temperature and ambient humidity (RH = 15–35%) at three cyclic rates (0.01, 0.25, and 3 Hz. [After Mostovoy and Ripling (197).]

TABLE 5.7   Fracture Toughness and Fatigue Thresholds in Adhesive Joints[a]

Adhesive	ΔG_{th}(lb/in.)[b]	G_{Ic}(lb/in.)	$G_{Ic}/\Delta G_{th}$
826/14.5M/340 (two-component epoxy)[c]	0.21	0.8	4
MRL-1 (modified nitrile epoxy on a polyester- fiber mat)	0.50	8.5	17
MRL-3 (unsupported nitrile-phenolic film)	1.6[d]	20	13
MRL-6 (modified epoxy impregnated into glass cloth)	≤0.3	5.8–13.5[e]	≥19–45
MRL-7 (modified epoxy, unsupported)	0.54[f]	10.4	19
MRL-8 (rubber-modified epoxy, unsupported)	0.45	15.7	35
MRL-2 (experimental nylon-epoxy, unsupported)	≤0.4	38.5	≥96

[a] From Mostovy and Ripling [197]. Unless otherwise specified, all tests were run at 3 Hz, with a sinusoidal wave form.

[b] Conversion: 1 lb/in. = 175 N/M.

[c] Normally used as a matrix in filament-wound composites, rather than as a structural adhesive.

[d] Independent of frequency from 0.01 to 3 Hz.

[e] Wide variations in G_{Ic}.

[f] At 5 Hz, but no effect of frequency between 0.01 and 3 Hz in water.

limiting value, an exponential crack growth law was found to hold until the maximum value of ΔG was noted just prior to catastrophic fracture:

$$da/dN = A\,\Delta G^n \tag{5.11}$$

[of course, Eq. (5.11) is the energy-based equivalent to the Paris relationship in terms of ΔK, Eq. (3.8)]. The validity of this relationship was established for seven different structural adhesives and tested under a variety of conditions (frequency, waveform, frequency, and environment); later studies by Marceau *et al.* [195] confirmed its validity for an unspecified adhesive. Interestingly, Eq. (5.11) held regardless of the locus of failure (center-of-bond or at the interface).

It is interesting to compare the values of fatigue threshold ΔG_{th} with values of the static fracture toughness G_{Ic} (see Table 5.7). First, values of ΔG_{th} are consistently much lower (by as much as two orders of magnitude)* than the values of G_{Ic}. Thus, as has been noted frequently for bulk polymers, fatigue loading is much more damaging than static loading. Second, there are wide variations from adhesive to adhesive. Although possible effects of differences in network structure were not studied, the variations still must reflect differences in the chemistry. In any case, the high *ratio* of G_{Ic} to ΔG_{th} is stated to compare favorably with the case of metals; the *absolute* value of ΔG_{th} is similar to values reported for aluminum.

Given that Eq. (5.11) holds over a reasonable range of ΔG, it is useful to compare the effects of frequency and environment with those observed in bulk plastics.

Effect of Frequency. As has been seen with bulk polymers, complex effects of frequency were observed. With a standard amine-cured epoxy (a resin matrix type rather than a structural adhesive), a significant effect of frequency was noted for $\Delta G_I \gtrsim 0.04$ lb/in. (7 N m). A common curve was found for frequencies of 0.25 and 0.01 Hz; its slope was nearly four times that for the curve obtained at 3 Hz. (It should be noted that the waveform was trapezoidal at 0.01 Hz and that immersion in water seemed to have no effect at the low frequencies.) Clearly, the higher the frequency, the lower the crack growth rate. Since the failure at the lower frequencies occurred at the interface rather than at the center of the bond, it is likely that the higher values of da/dN were due to the attack of water on the adherend–primer adhesive bond. Therefore the effect of frequency may well be related to time-

---

* It should be noted that in some cases the values estimated for ΔG_{th} were not explicitly determined, but were inferred from the general pattern of data. Hence there may be significant error in the absolute values of ΔG_{th}; however, the trends noted should not be affected.

dependent chemical effects rather than to the effect of frequency *per se* on material behavior.

Quite different behavior was noted in two other cases. With a standard nitrile–phenolic film-type adhesive (Fig. 5.53), no significant effect of frequency was observed when the frequency was changed by a factor of 300. With a modified epoxy adhesive, no effect of frequency was observed on increasing the frequency from 0.25 to 3 Hz; however, at 5 Hz, da/dN tended to be lower than at 3 Hz. Behavior of this type has been seen with bulk polymers (see Section 3.4), and attributed to the onset of crack blunting due to hysteretic heating at the higher frequency.

An effect of frequency on fatigue life of lap-shear adhesive joints has also been observed by Marceau *et al.* [195], though the adhesive was not specified (Marceau *et al.* also confirmed the sigmoidal nature of plots of da/dN versus ΔG.) The fatigue life and the values of da/dN at constant ΔG at 2×10^{-4} Hz were far less than at 30 Hz. However, since the load profiles were quite different, the low frequency being trapezoidal and the high frequency sinusoidal, precise comparison is not possible.

Thus, although the data do not permit more extensive analysis, it is seen that two types of frequency behavior observed in bulk polymers are also seen in adhesive joints: (1) the case of no effect of frequency, and (2) the case of lower crack growth rates at a higher frequency. Unfortunately, direct comparison with the results for bulk polymers discussed in Section 3.4 is not possible at this time due to differences in environment and possible differences between bulk and adhesive-layer properties.

Environment and Temperature Effects. Under static loading conditions, Mostovoy and Ripling [197] found that the general effect of exposure to water was small for a group of four adhesives tested at temperatures close to ambient, although some exhibited a slight toughening at somewhat elevated temperatures (e.g. at 38°C). However, at higher temperatures, G_{Ic} was invariably decreased as was G_{Iscc} (the value of G_I below which cracking did not occur in the presence of water).

In fatigue, the effects of water exposure considerably varied. With a matrix-type epoxy, failure tended to occur at the interface, and FCP rates were high, though no higher when liquid water was used, presumably because water vapor was already attacking the interface in the ambient control tests. With the nitrile phenolic, a complex interaction between load and environment was noted. At 3 Hz, it was seen that water improved fatigue performance at low ΔG_I, but had little effect at high values; in other words, the curve in liquid water had a higher slope than that for ambient conditions. In contrast, Marceau *et al.* [195] found little effect of moisture in their experiments. (For further discussion of moisture effects on various aspects of failure see Section 3.6 and Manson and Sperling [3, Chap. 12] and Bascom and Cottington [191].)

In the Mostovoy–Ripling study [197], the effect of one temperature other than ambient was examined in one adhesive system, MRL-3. While the shape of the fatigue curve was similar to that at room temperature, the curve was shifted vertically, so that at 56°C, instead of 24°C, the crack propagation rates were higher (by about an order of magnitude) at a given value of ΔG_I, and the value of ΔG_{th} was reduced by a factor of about 4. Increases in da/dN at constant ΔG_I (about twofold) were noted by Marceau *et al.* [195] on going from 24 to 60°C. A corresponding decrease in fatigue life was also noticed in S–N tests.

The locus of failure apparently depends on the test conditions employed. With all but one of the adhesives studied by Mostovoy and Ripling [197], the failure at room temperature was center of bond. With MRL-3, at 56°C, however, the failure tended to occur between the primer and the adhesive, behavior analogous to what is seen with stress-corrosion cracking induced by water. In the studies by Marceau *et al.* [195], failure also occurred close to the primer layer in the lap-shear specimens tested at 30 Hz. At lower frequencies (and with sustained loads), the failure was center-of-bond. It may well be that creep plays a dominant role at the lower frequencies.

Implications for Design. Even though the studies mentioned have been limited in scope, it is desirable that loads above G_{th} should generally be avoided. Since the correlation between G_{th} and G_{Ic} is crude, with significant exceptions, it will be necessary to measure G_{th} for each system under consideration, and at frequencies and in environments anticipated in service. It should also be recognized that waveform may be important, and that actual loadings in service commonly involve a combination of modes I and II. Thus it would be desirable to extend studies of mixed-mode loading (see, for example, Bascom [199]) to include fatigue phenomena.

In any case, studies of the type discussed should be very useful in analyzing fatigue in adhesive joints, in the selection of materials, and in design. The question of loading patterns in service is also important (see, for example, Wilkins *et al.* [200]); simulation of realistic load histories is clearly desirable.

Relationship to Bulk Polymers and Metals. Comparison of the fatigue behavior of adhesive joints with that of bulk polymers and metals, for example, aluminum, has been made by Mostovoy and Ripling [197], and by Marceau *et al.* [195]. First, with a few exceptions, the relationship between fatigue crack growth rate (with load control) and fracture toughness reported by Manson and Hertzberg [6] for bulk polymers was confirmed (Fig. 5.54). Second, a comparison of adhesive joints with bulk polymers (Fig. 5.55)

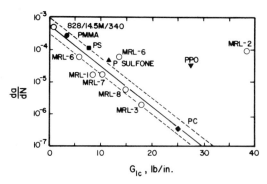

Fig. 5.54 Comparison of G_{Ic} and da/dN in fatigue at a given ΔG_I level for seven adhesives. Data for structural plastic materials from Ref. 6. (ΔK at which da/dN is evaluated is 1 ksi - $\sqrt{\text{in.}}$, $\Delta G = 2.22$ lb/in., $E = 0.45 \times 10^6$ psi for the adhesives.) Equation line and biased standard error of the estimate(s) is least-squares fit to ten data points for adhesives (large circles) and plastics (small plotting symbols). Fatigue exposure is 3 Hz at room temperature (RT) and ambient humidity (RH = 10–40%). Conversions: 1 lb/in. = 175 N/m; 1 ksi $\sqrt{\text{in.}}$ = 1.1 MPa·m$^{1/2}$. [After Mostovoy and Ripling (197).]

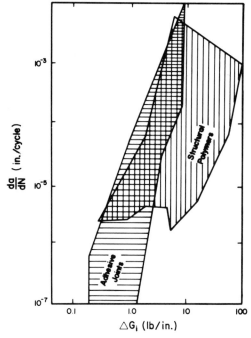

Fig. 5.55 Comparison of the range of da/dN versus ΔG_I values obtained for structural polymers and adhesive joints exposed to room temperature (RT), ambient humidity, and fatigue at frequencies between 1 and 10 Hz. Boundary lines indicate the range where data is available; e.g., there is no ΔG_{th} data for bulk polymers. Conversion: 1 lb/in. = 175 N/m. [After Mostovoy and Ripling (197).]

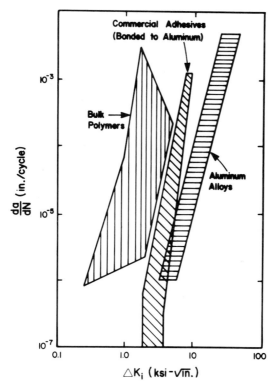

Fig. 5.56 Comparison of *da/dN* versus ΔK_I plot, based on load carrying capacity at room temperature (RT), for fatigue of bulk polymers, aluminum alloys, and commercial adhesives bonded to aluminum. Conversion: 1 ksi $\sqrt{in}$. = 1.1 MPa $\cdot$ m$^{1/2}$. [After Mostovoy and Ripling (197).]

shows that the better of the bulk polymers have greater fatigue resistance than the adhesive joints, at least in terms of the strain energy release associated with a given crack growth rate. Thus there is room for improvement in the fatigue behavior of structural adhesives. Third, a comparison of structural adhesives with aluminum alloys (Fig. 5.56) shows that, in terms of load-carrying capacity (as implied by plotting *da/dN* versus ΔK), at least seven adhesively bonded aluminum joints were more fatigue-resistant than the best of the bulk polymers (see also Fig. 3.40) and comparable to some typical aluminum alloys themselves at low values of ΔK. Similar results were reported by Marceau *et al.* [195]. Comparisons of this kind can be helpful in putting the behavior of adhesive joints into proper perspective. While there is much concern over the reliability of adhesive joints, and frequently suspicion of adhesives generally, fatigue crack growth rates in well-made joints do seem to fall at least within an order of magnitude of those for aluminum alloys at the highest loadings. Of course, as pointed out above, effects of temperature and more complex stress states may significantly alter the relative behavior.

5.4.2 Polymer Concrete Composites

Recently great interest has developed in several types of polymer–concrete composites for engineering applications. These include three principal types:

(1)   polymer-impregnated concrete (PIC) in which a previously fabricated porous solid (typically, portland cement concrete) is impregnated with a monomer or prepolymer which is subsequently polymerized in situ;

(2)   polymer-portland-cement-concrete (PPCC) in which a polymer (typically in latex form) is combined with a portland cement mix prior to curing of the cement phase; and

(3)   polymer concrete (PC)* in which a polymer is used as a binder for an aggregate such as sand.

However, in spite of the fact that loadings of such materials used in structural or other applications are often cyclic in nature, little has been reported on fatigue behavior.

With PIC, impregnation with < 10 wt% PMMA improves both compressive and tensile strengths severalfold; the improvement has been attributed to several factors including the reduction of porosity, specific increases in fracture energy, and the reduction of stress concentrations in the brittle matrix [201]. These improvements in static strength may well be paralleled by significant improvements in fatigue response. As shown in Table 5.8, in preliminary experiments on portland cement mortar (in flexure at 40 Hz) the incorporation of ~ 9 wt% PMMA permitted cycling at loads five times higher than was possible with control specimens [203]. Even at these higher loads, fatigue lives were increased severalfold; about half the specimens had not broken at 10^6 cycles. The improvement in fatigue life may be attributed to the general factors just mentioned with respect to static strength, and, specifically, to improvements in the strength of the sand–matrix bond (a conclusion confirmed by microscopic examination). Fatigue tests on larger concrete bars have also been reported by Kuckacka *et al.* [204]. Specimens underwent 10^6 test cycles at 77% of the ultimate strength without failure, a finding in agreement with the results just discussed. It is interesting to note that, due to the continuous nature of the capillary pore

---

* This abbreviation, recommended by the American Concrete Institute, should not be confused with the standard abbreviation used by polymer scientists and engineers for polycarbonate.

TABLE 5.8 Fatigue Behavior of Polymer-Impregnated Cement (PIC) Mortar[a,b]

Property	PIC	Control
Ultimate flexural strength (UFS) (psi)[c]	5300	900
Upper stress level (75% of UFS) (psi)	3960	682
Lower stress level (25% of UFS) (psi)	1320	228
Frequency (Hz)	40	40
Number of cycles to failure:		
specimen 1	236,100	116,000
specimen 2	190,700	90,100
specimen 3	370,000	77,700
specimen 4	900,000[c]	254,300
specimen 5	900,000[c]	188,600

[a] Gebauer and Manson [203].

[b] Containing 8.6 wt% poly(methyl methacrylate).

[c] Conversion: 1000 psi = 6.9 MPa.

[d] Fatigue failure was not obtained after 900,000 cycles.

system that is impregnated, PIC composites are essentially polymer-fiber-reinforced concretes, the fibers being of microscopic dimensions. Also, at least with PMMA, there is evidence for strong interfacial bonding [63, 106].

While with PPCC, fatigue data do not appear to be available, preliminary information has been developed for at least some PC systems, primarily because some PCs are of interest as orthopedic or dental cements for bones or teeth. While these may formally be classed as PCs, they are arbitrarily discussed as particulate-reinforced systems because the polymer concentration ($\sim 40\%$) is higher than is typical for PCs.

If designers are to use polymer concrete systems extensively as structural materials, much more knowledge about fatigue will be necessary. As discussed in Section 5.2, interfacial effects are bound to be important in determining long term strength under cyclic loading.

APPENDIX

To aid in defining possible modes of failure in composites, Table 1 and Figs. 1–7, inclusive, are provided. Since it is sometimes difficult for the novice to see what the expert sees in his photomicrographs, these should be helpful in clarifying the interpretation of failure.

The thorough enumeration by Chamis [87] is especially useful for this purpose.

TABLE 1   Possible Failure Modes for a Ply in a Continuous-Fiber Composite[a]

Mode of loading	Failure response	Mode of loading	Failure response

Longitudinal tensile (Fig. 1)	(a) Brittle
	(b) Brittle with filament pullout
	(c) Staggered failure brittle with filament pullout and
	  (1) interfilament matrix shear
	  (2) constituent debonding, i.e., the matrix breaks away from the filaments
		Transverse tensile (Fig. 5)	(a) Matrix tensile failure
			(b) Matrix tensile failure, constituent debonding, and/or fiber splitting

		Transverse compression (Fig. 6)	(a) Matrix compressive failure (b) Matrix shear failure (c) Matrix shear failure constituent debonding, and/or fiber crushing
Longitudinal compression (Figs. 2, 3, and 4)	(a) Filament microbuckling, matrix still elastic (b) Matrix yield followed by filament microbuckling (c) Constituent debonding followed by filament microbuckling (d) Panel microbuckling (e) Shear failure (f) Ply separation by transverse tension through the thickness (transverse tensile splitting)	In-plane shear (intralaminar shear) (Fig. 7)	(a) Matrix shear failure known as intralaminar, also called interlaminar, shear (b) Matrix shear failure and constituent debonding (c) Constituent debonding

[a] From Chamis [87].

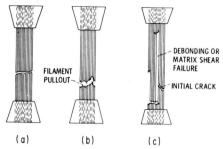

Fig. A.1 Longitudinal tensile failure modes: (a) brittle, (b) brittle with filament pullout; (c) irregular (from Chamis [87]).

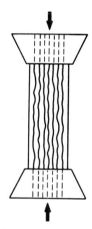

Fig. A.2 Longitudinal compressive failure mode by microbuckling (from Chamis [87]).

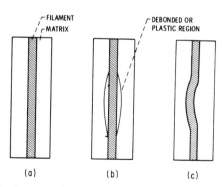

Fig. A.3 Longitudinal compressive failure mode by constituent debonding: (a) constituents intact; (b) constituents debonded or matrix yielded; (c) buckled filament (from Chamis [87]).

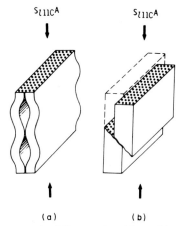

Fig. A.4 Longitudinal compressive failure modes: (a) panel buckling or splitting; (b) shear failure (A is the cross-sectional area) (from Chamis [87]).

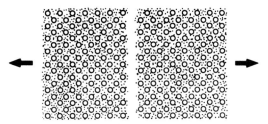

Fig. A.5 Transverse failure mode (600 ×): schema showing transverse failure.

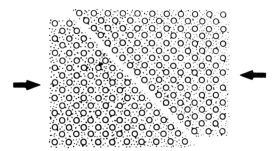

Fig. A.6 Schema showing transverse compressive failure mode (from Chamis [87]).

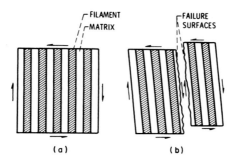

Fig. A.7 Schema showing shear failure modes (in-plane, intralaminar): (a) unfailed; (b) failed (from Chamis [87]).

REFERENCES

[1] A. Kelly, "Strong Solids." Oxford Univ. Press, London and New York, 1966.
[2] L. J. Broutman and R. H. Krock (eds.), "Composite Materials." 8 Vols. Academic Press, New York, 1974.
[3] J. A. Manson and L. H. Sperling, "Polymer Blends and Composites." Plenum Press, New York, 1976.
[4] J. E. Gordon, "The New Science of Strong Materials." 2nd ed., Penguin Books, Harmondsworth, England, 1976.
[5] J. R. Beatty, *Rubber Chem. Technol.* **37**, 1341 (1964).
[6] J. A. Manson and R. W. Hertzberg, *Crit. Rev. Macromol. Sci.* **1**, 433 (1973).
[7] C. B. Bucknall, "Toughened Plastics." Applied Science Publishers, London, 1977.
[8] I. I. Ostromislensky (to Naugatuck Chem.), U. S. Patent 1,613,673 (January 11, 1927).
[9] "Modern Plastics Encyclopedia." McGraw-Hill, New York, 1977.
[10] R. P. Kambour, *J. Polym. Sci. Macromol. Rev.* **7**, 1 (1973).
[11] J. Mann and G. R. Williamson, *in* "The Physics of Glassy Polymers." (R. N. Haward, ed.), p. 454. Wiley, New York, 1973.
[12] H. Keskkula, A. E. Platt, and R. F. Boyer, *in* "Encyclopedia of Chemical Technology." 2nd ed., Vol. 19, p. 85. Wiley (Interscience), New York, 1969.
[13] P. Beahan, A. Thomas, and M. Bevis, *J. Mat. Sci.* **11**, 1207 (1976).
[14] C. B. Bucknall and I. C. Drinkwater, *J. Mat. Sci.* **8**, 1800 (1973).
[15] C. B. Bucknall, D. Clayton, and W. Keast, *J. Mat. Sci.* **7**, 1443 (1972).
[16] C. B. Bucknall, I. C. Drinkwater, and W. Keast, *Polymer* **13**, 115 (1972).
[17] G. Gigna, *J. Appl. Polym. Sci.* **14**, 1781 (1970).
[17a] J. D. Ferry, "Viscoelastic Properties of Polymers." 2nd ed., Wiley, New York, 1970.
[18] E. R. Wagner and L. M. Robeson, *Rubber Chem. Technol.* **43**, 1129 (1970).
[19] L. Cessna, *Am. Chem. Soc. Polym. Preprint* **15**(1), 229 (1974).
[20] C. B. Bucknall and W. W. Stevens, *in* "Toughening of Plastics," paper 24. Plastics and Rubber Institute, London, 1978.
[21] H. Bartesch and D. R. G. Williams, *J. Appl. Polym. Sci.* **22**, 467 (1978).
[22] C. B. Bucknall, *Brit. Plast.* **40**(11), 118; **84** (1967).
[23] R. P. Kambour and R. W. Kopp, *J. Polym. Sci. A* **27**, 183 (1969).
[24] C. B. Bucknall, *J. Mat.* **4**, 214 (1969).
[25] R. Murukami, N. Shin, N. Kusomoto, and Y. Motozato, *Kobunshi Robunshu* **33**, 107 (1976).

[26] P. Beardmore and S. Rabinowitz, *Appl. Polym. Symp.* **24**, 25 (1974).

[27] G. Menges, *Proc. Ann. Tech. Conf., 19th* p. 180. Soc. Plastic Eng. (1973).

[28] G. Menges and E. Wiegand, *Proc. Ann. Tech. Conf., 21st* p. 469. Soc. Plast. Eng. (1975).

[29] R. L. Thorkildsen, *in* "Engineering Design for Plastics" (E. Baer, ed.), p. 279. Van Nostrand-Reinhold, Princeton, New Jersey, 1964.

[30] C. B. Bucknall, K. V. Gotham, and P. I. Vincent, *in* "Polymer Science. A Materials Handbook" (A. D. Jenkins, ed.), Vol. 1, Chapter 10. American Elsevier, New York, 1972.

[31] M. D. Skibo, J. Janiszewski, S. L. Kim, R. W. Hertzberg, and J. A. Manson, *Proc. Ann. Tech. Meeting, 36th, SPE, Washington, D. C.* p. 304 (1978).

[32] M. D. Skibo, J. Janiszewski, R. W. Hertzberg, and J. A. Manson, *Proc. Int. Conf. Toughening Plast. London* Paper 25 (July 1978).

[33] M. D. Skibo, J. A. Manson, R. W. Hertzberg, and E. A. Collins, *in* "Durability of Macromolecular Materials" (R. K. Eby, ed.), ACS Symposium Series No. 95; American Chemical Society: Washington, D.C., 1979, p. 311.

[34] M. D. Skibo, R. W. Hertzberg, and J. A. Manson, *Proc. Conf. Deformat. Fracture.* Plastics and Rubber Institute, London, 1979.

[35] M. D. Skibo, R. W. Hertzberg, and J. A. Manson, *J. Mat. Sci.* **14**, 2482 (1979).

[36] M. D. Skibo, R. W. Hertzberg, and J. A. Manson, *J. Mat. Sci.* **11**, 479 (1976).

[37] J. P. Elinck, J. -C. Bauwens, and G. Homès, *Int. J. Fract. Mech.* **7**, 227 (1971).

[38] N. J. Mills and N. Walker, *Polymer* **17**, 335 (1976).

[39] M. Skibo, R. W. Hertzberg, J. A. Manson, and E. A. Collins, *Proc. Int. Conf. PVC, 2nd, Lyon* p. 233, (1976).

[40] J. Murray and D. Hull, *Polymer* **10**, 451 (1969).

[41] M. D. Skibo, R. W. Hertzberg, and J. A. Manson, *Proc. Int. Conf. Fracture, 4th, Waterloo, Canada* p. 1127 (1977).

[42] R. W. Hertzberg, J. A. Manson, and M. D. Skibo, *Polymer* **19**, 359 (1978).

[43] R. Attermo and G. Ostberg, *Int. J. Fract. Mech.* **7**, 122 (1971).

[44] E. A. Flexman, *Proc. Int. Conf. Toughening Plast. London,* Paper 14 (July 1978).

[45] E. Krokosky, *in* "Modern Composite Materials" (L. J. Broutman and R. H. Krock, eds.), Chapter 4. Addison-Wesley, Reading, Massachusetts, 1967.

[46] C. B. Bucknall, *in* "Failure of Polymers" (*Adv. Polym. Sci.*), Vol. 27, p. 121. Springer-Verlag, Berlin and New York, 1978.

[47] Y. S. Lipatov and L. M. Sergeva, "Adsorption of Polymers." Wiley, New York, 1974.

[48] R. P. Kusy and D. T. Turner, *J. Dental Res.* **53**, 520 (1974).

[49] P. W. R. Beaumont and R. J. Young, *J. Mat. Sci.* **10**, 1334 (1975).

[50] J. M. Powers, J. C. Roberts, and R. G. Craig, *J. Dental Res.* **55**, 432 (1976).

[51] A. Grant, *Brit. Polym. J.* **10**, 241 (1976).

[52] J. Lankford, W. J. Astleford, and M. A. Asher, *J. Mat. Sci.* **11**, 1624 (1976).

[53] F. F. Koblitz, V. R. Luna, J. F. Glenn, K. L. DeVries, and R. A. Draughn, *Org. Coat. Plast. Chem.* **38**(1), 322 (1978).

[54] G. P. Pearson and D. F. Jones, *Brit. Polym. J.* **10**, 256 (1978).

[55] T. A. Freitag and S. L. Cannon, *J. Biomed. Mat. Res.* **11**, 609 (1977).

[56] T. A. Freitag and S. L. Cannon, *J. Biomed. Mat. Res.* **13**, 343 (1979).

[57] C. F. Stark, *J. Biomed. Mat. Res.* **13**, 341 (1979).

[58] J. Aoki, *Porima Daijesuto* **29**, 25 (1977); see *Chem. Abstr.* **89**, 90497 (1978).

[59] J. Aoki, *Raba Daijesuto* **29**, 37 (1977); see *Chem. Abstr.* **88**, 153339 (1978).

[60] L. E. Nielsen, *J. Composite Mater.* **9**, 149 (1975).

[61] R. M. Pilliar, R. Blackwell, I. Macnab, and H. U. Cameron, *J. Biomed. Mater. Res.* **10**, 893 (1976).

[62] R. M. Pilliar, W. J. Bratina, and R. A. Blackwell, ASTM STP 636, p. 206 (1977).

[63] M. J. Marmo, M. A. Mostafa, H. Jinnai, F. M. Fowkes, and J. A. Manson, *Ind. Eng. Chem. Prod. Res. Dev.* **15**, 206 (1976).

[64] F. M. Fowkes and M. Mostafa, *Ind. Eng. Chem. Prod. Res. Dev.* **17**, 3 (1978).

[65] G. P. Sendeckyj (ed.), "Mechanics of Composite Systems," Vol. 2, Composite Materials. Academic Press, New York, 1974.

[66] L. J. Broutman and R. H. Krock, "Modern Composite Materials." Addison-Wesley, Reading, Massachusetts, 1967.

[67] H. C. Corten, *in* "Modern Composite Materials" (L. J. Broutman and R. H. Krock, eds.), Chapter 2. Addison-Wesley, Reading, Massachusetts, 1969.

[68] D. Dew-Hughes and J. L. Way, *Composites* **4**, 167 (1973).

[69] M. J. Owen, *in* "Fracture and Fatigue," Vol. 5, Composite Materials, Chapters 7 and 8, Academic Press, New York, 1974.

[70] B. Harris, *Composites* **8**, 214 (1977).

[71] K. L. Reifsnider, *Int. J. Fracture* (1979) (in press).

[72] "Fatigue of Composite Materials," STP 569. American Society for Testing and Materials, Philadelphia, Pennsylvania, 1973.

[73] "Fatigue of Filamentary Composite Materials," STP 636. American Society for Testing and Materials, Philadelphia, Pennsylvania, 1973.

[74] K. L. Reifsnider and K. N. Lauraitis, (eds.), "Fatigue of Fibrous Composite Materials," STP XXX, American Society for Testing and Materials, Philadelphia, Pennsylvania, 1980 (in press).

[75] "Fatigue in Composites" (*Proc. One-Day Symp.*). Institute of Physics, London (November 1973).

[76] "Fatigue Crack Growth under Spectrum Loads," STP 595. American Society for Testing and Materials, Philadelphia, Pennsylvania, 1976.

[77] "Environmental Effects on Advanced Composite Materials," STP 602. American Society for Testing and Materials, Philadelphia, Pennsylvania, 1976.

[78] "Use of Computers in the Fatigue Laboratory," STP 613. American Society for Testing and Materials, Philadelphia, Pennsylvania, 1976.

[79] "Composite Materials: Testing and Design" (Fourth Conference), STP 617. American Society for Testing and Materials,

[80] "Cyclic Stress-strain and Plastic Deformation Aspects of Fatigue Crack Growth," STP 637. American Society for Testing and Materials, Philadelphia, Pennsylvania, 1977.

[81] "Advanced Composite Materials—Environmental Effects," STP 658. American Society for Testing and Materials, Philadelphia, Pennsylvania, 1978.

[82] "Composite Materials: Testing and Design" (Fifth Conference), STP 674. American Society for Testing and Materials, Philadelphia, Pennsylvania, 1979.

[83] J. N. Fleck and R. L. Mehan (eds.), "Failure Modes in Composites." The Metallurgical Society of AIME, New York, 1974.

[84] A. S. Argon *in* "Fracture and Fatigue" (L. J. Broutman, ed.), Vol. 5, Composite Materials, p. 154. Academic Press, New York, 1974.

[85] K. H. Boller, *Mod. Plast.* **34**, 163 (1957); **41**, 145 (1964).

[86] L. J. Broutman (ed.), "Fracture and Fatigue," Vol. 5, Composite Materials. Academic Press, New York, 1974.

[87] C. C. Chamis, *in* "Fracture and Fatigue" (L. J. Broutman, ed.). Vol. 5, Composite Materials, p. 94. Academic Press, New York, 1974.

[88] J. M. Lifschitz, *in* "Fracture and Fatigue" (L. J. Broutman, ed.). Vol. 5, Composite Materials, Chapter 6. Academic Press, New York, 1974.

[89] B. Parkyn (ed.), "Glass Reinforced Plastics." CRC Press, Cleveland, Ohio, 1970.

[90] R. J. Schwartz and H. S. Schwartz, "Fundamental Aspects of Fiber-Reinforced Plastic Composites." Wiley (Interscience), New York, 1968.

[91] S. W. Tsai, J. C. Halpin, and N. J. Pagano (eds.), "Composite Materials Workshop." Technomic Publ. Stamford, Connecticut, 1969.

[92] E. M. Wu, in "Fracture and Fatigue" (L. J. Broutman, ed.). Vol. 5, Composite Materials, Chapter 3. Academic Press, New York, 1974.

[93] M. R. Piggott, *J. Mat. Sci.* **9**, 494 (1974).

[94] T. U. Marston, A. G. Atkins, and D. K. Felbeck, *J. Mat. Sci.* **9**, 447 (1974).

[95] Anonymous, in "Modern Plastics Encyclopedia," pp. 124, 172. McGraw-Hill, New York, 1978.

[96] Anonymous, *Mod. Plast.* **54**(2), 38 (1977).

[97] R. E. Lavengood and L. E. Gulbransen, *Polym. Eng. Sci.* **9**, 365 (1969).

[98] J. W. Dally and L. J. Broutman, *J. Composite Mat.* **1**, 424 (1967).

[99] L. C. Cessna, J. A. Levens, and J. B. Thomson, *Polym. Eng. Sci.* **9**, 339 (1969).

[100] R. G. Shaver and E. F. Abrams, *Proc. Ann. Tech. Conf., Soc. Plast. Eng.* **17**, 378 (1971)

[101] J. W. Dally and D. H. Carrillo, *Polym. Eng. Sci.* **9**, 434 (1969).

[102] J. Soldatos and R. Burhans, *Ind. Eng. Chem. Prod. Res. Dev.* **9**, 296 (1970).

[103] M. J. Owen and R. G. Rose, *Plast. Polym.* **40**, 325 (1972).

[104] M. Kodama, *Kobunshi Robunshu* **31**, 415 (1974); **32**, 66 (1975).

[105] M. Kodama, *J. Appl. Polym. Sci.* **20**, 2853 (1976).

[106] Y. N. Liu, J. A. Manson, J. W. Vanderhoff, and W. F. Chen, *Polym. Eng. Sci.* **17**, 325 (1977).

[107] M. J. Owen and R. Dukes, *J. Strain Anal.* **2**, 272 (1967).

[108] M. J. Owen, T. R. Smith, and R. Dukes, *Plast. Polym.* **37**, 227 (1969).

[109] J. Theberge, B. Arkles, and R. Robinson, *Ind. Eng. Chem. Prod. Res. Dev.* **15**, 100 (1976).

[110] J. Y. Lomax and J. T. O'Rourke, *Proc. Ann. Tech. Conf. Soc. Plast. Eng., 21st* Paper X-5 (1965).

[111] T. Misaki and J. Kishi, *J. Appl. Polym. Sci.* **22**, 2063 (1978).

[112] A. T. DiBenedetto, G. Salee, and R. Hlavacek, *Polym. Eng. Sci.* **15**, 242 (1975).

[113] A. T. DiBenedetto and G. Salee, Tech. Rep. AMMRC CTR 77-12 (March 1977).

[114] A. T. DiBenedetto and G. Salee, *Polym. Eng. Sci.* **18**, 634 (1978).

[115] A. T. DiBenedetto and G. Salee, *Polym. Eng. Sci.* **19**, 512 (1979).

[116] J. C. Halpin, K. L. Jerina, and T. A. Johnson, Tech. Rep. AFML TR-73-289 (December 1972).

[117] P. A. Thornton, *J. Composite Mater.* **6**, 147 (1972).

[118] M. J. Owen and P. T. Bishop, *J. Phys. D: Appl. Phys.* **7**, 1214 (1974).

[119] A. W. Holdsworth, S. Morris, and M. J. Owen, *J. Phys. D: Appl. Phys.* **7**, 2036 (1974).

[120] S. L. Kim, M. D. Skibo, J. A. Manson, R. W. Hertzberg, and J. Janiszewski, *Polym. Eng. Sci.* **18**, 1093 (1978).

[121] P. Bretz, R. W. Hertzberg, and J. A. Manson (1979) ACS Symposium Series, (in press).

[122] F. Erdogan, in "Crack Propagation Theories—An Advanced Treatise," Chapter 2. Academic Press, New York, 1968.

[123] M. Kitagawa and K. Motomura, *J. Polym. Sci.* **12** (1974).

[124] L. J. Broutman, in "Modern Composite Materials" (L. J. Broutman and R. H. Krock, eds.), Chapter 13. Addison-Wesley, Reading, Massachusetts, 1969.

[125] B. H. Jones, in "Modern Plastics Encyclopedia," p. 368. McGraw-Hill, New York, 1978–1979.

[126] F. J. McGarry, in "Fundamental Aspects of Fibre Reinforced Plastic Composites" (R. T. Schwartz and H. S. Schwartz, eds.), p. 63. Wiley (Interscience), New York, 1968.

[127] L. J. Broutman and S. Sahu, *Ann. Tech. Conf. SPI, 24th* Paper 11D (1969).

[128] M. Fuwa, B. Harris, and A. R. Bunsell, *J. Phys. D: Appl. Phys.* **8**, 1460 (1975).

[129] J. J. Nevadunsky, J. J. Lucas, and M. J. Salkind, *J. Composite Mat.* **9**, 394 (1975).

[130] J. E. Schwabe, ASTM STP 580, p. 396 (1975).

[131] R. D. Adams, D. Walton, J. E. Flitcroft, and D. Short, ASTM STP 580, p. 159 (1975).

[132] W. W. Stinchcomb, K. L. Reifsnider, L. A. Marcus, and R. S. Williams, ASTM STP 569, p. 115 (1975).

[133] A. T. DiBenedetto, J. V. Gauchel, R. L. Thomas and J. W. Barlow, *J. Mat.* **7**, 211 (1972).

[134] F. H. Chang, D. E. Gordon, and A. H. Gardner, ASTM STP 636, p. 57 (1977).

[135] G. L. Roderick and J. D. Whitcomb, ASTM STP 636, p. 73 (1977).

[136] G. P. Sendeckyj and H. D. Stalnaker, ASTM STP 617, p. 39 (1977).

[137] D. T. Hayford and E. G. Henneke, II, ASTM STP 674, 184 (1979).

[137a] C. T. Sun and W. S. Chan, ASTM STP 674, 418 (1979).

[138] J. H. Williams and S. S. Lee, *J. Composite Mat.* **12**, 348 (1978).

[139] H. C. Kim and L. J. Ebert, *J. Composite Mat.* **10**, 139 (1978).

[140] M. J. Owen and S. Morris, *Int. Conf. Carbon Fibres, Their Composites Appl.* Paper 51. Plastics Inst., London, 1971.

[141] P. W. R. Beaumont and B. Harris, *Int. Conf. Carbon Fibres, Their Composites Appl.* Paper 49. Plastics Inst., London, 1971.

[142] D. G. Fesko, *Polym. Eng. Sci.* **17**, 242 (1977).

[143] S. Morris, Ph.D. Thesis, Univ. of Nottingham, England (1970), quoted in Owen [69, Chapter 8].

[144] D. Schutz and J. J. Gerharz, *Composites* **8**, 245 (1977).

[145] C. K. H. Dharan, *in* "Failure Modes in Composites II" (J. N. Fleck and R. L. Mehan, eds.), p. 144. The Metallurgical Society of AIME, New York, 1974.

[146] C. K. H. Dharan, *J. Mat. Sci.* **10**, 1665 (1975).

[147] C. K. H. Dharan, *Proc. 1975 Int. Conf. Composite Mat.* Vol. 1, p. 830. The Metallurigical Society of AIME, New York, 1976.

[148] C. K. H. Dharan, ASTM STP 569, p. 171 (1975).

[149] D. C. Phillips and J. M. Scott, *Composites* **8**, 233 (1977).

[150] L. G. Bevan, *Composites* **8**, 227 (1977).

[150a] S. V. Ramani and D. P. Williams, ASTM STP 636, 27 (1977).

[151] K. E. Hofer, Jr., L. C. Bennett, and M. Stander, ASTM STP 636, p. 103 (1977).

[152] J. T. Ryder and E. K. Walker, ASTM STP 636, p. 3 (1977).

[153] D. F. Sims and V. H. Brogdon, ASTM STP 636 p. 185 (1977).

[154] C. T. Sun and G. L. Roderick, ASTM STP 636, p. 89 (1977).

[155] J. B. Sturgeon, *Composites* **8**, 221 (1977).

[156] J. V. Gauchel, I. Steg, and J. E. Cowling, ASTM STP 569, p. 45 (1975).

[157] E. W. Smith and K. J. Pascoe, *Composites* **8**, 237 (1977).

[158] S. V. Kulkarni, P. V. McLaughlin, Jr., R. B. Pipes, and B. W. Rosen, ASTM STP 617, p. 70 (1977).

[159] L. H. Miner, R. A. Wolffe, and C. H. Zweben, ASTM STP 580, p. 549 (1975).

[160] H. Miyairi, M. Nagai, and A. Murumatsu, *Proc. Jpn. Congr. Mat. Res.-Nonmetall. Mat., 19th* p. 189 (1976).

[161] J. C. Halpin, K. L. Jerina, and T. A. Johnson, ASTM STP 521, p. 5 (1973).

[162] H. T. Hahn, ASTM STP 674, p. 383 (1979).

[163] H. T. Hahn and R. Y. Kim, *J. Composite Mat.* **9**, 297 (1975).

[164] P. C. Chou and R. Croman, *J. Composite Mat.* **12**, 177 (1978).

[164a] P. C. Chou and R. Croman, ASTM STP 674, 431 (1979).

[165] J. Awerbuch and H. T. Hahn, ASTM STP 636, p. 248 (1977).

[166] J. N. Yang and M. D. Liu, *J. Composite Mater.* **11**, 176 (1977).

[167] H. Nakayasu, Z. Mackawa, T. Fujii, and K. Mizuwaka, *Proc. Jpn. Congr. Mat. Res.-Nonmetall. Mat., 19th* p. 195 (1976).

[167b] D. G. Harlow, ASTM STP 674, p. 484 (1978).

[168] P. V. McLaughlin, Jr., S. V. Kulkarni, S. N. Huang, and B. W. Rosen, Fatigue of Notched Fiber Composite Laminates Part I: Analytical Model. Materials Science Corporation, MSC/TFR/7501 (March 1975).

[169] J. N. Yang and D. L. Jones, *J. Composite Mater.* **12**, 371 (1978).

[170] J. N. Yang, *J. Composite Mater.* **12**, 19 (1978).

[170a] S. L. Phoenix, ASTM STP 674, 455 (1979).

[171] J. N. Yang and D. L. Jones, *Proc. AIAA/ASME/ASCE/ASH Structures, Structural Dynam., Mat. Conf., 20th* p. 232 (1979).

[172] J. H. Underwood and D. P. Kendall, *Proc. 1975 Int. Conf. Composite Mat.* **2**, 1122 (1976).

[173] R. B. Pipes, S. V. Kulkarni, and P. V. McLaughlin, *Mat. Sci. Eng.* **30**, 113 (1977).

[174] T. R. Porter, ASTM STP 636, p. 152 (1977).

[175] C. Y. Kam, paper presented, *Symp. Fatigue Fibrous Composite Mat., ASTM, San Francisco, May 22–23* (1979).

[176] J. W. Davis, J. A. McCarthy, and J. N. Schurb, *Mat. Design Eng.* **60**, 87 (1964).

[177] H. F. Brinson and Y. T. Yeow, ASTM STP 617, p. 18 (1977).

[178] M. J. Owen and P. T. Bishop, *J. Phys. D* **6**, 2057 (1973).

[179] M. S. Rosenfeld and C. W. Gause, *Symp. Fatigue Fibrous Composite Mat., ASTM, San Francisco, May 22–23* (1979).

[180] R. W. Walter, R. W. Johnson, R. R. Jone, and J. E. McCarty, ASTM STP 636, p. 228 (1977).

[181] W. S. Carswell, *Composites* **8**, 251 (1977).

[182] E. F. Kanninen, E. F. Rybicki, and W. I. Griffith, ASTM STP 617, p. 53 (1977).

[183] J. F. Mandell and U. Meier, ASTM STP 569, p. 28 (1975).

[184] M. D. Campbell and B. W. Cherry, *in* "Fracture Mechanics and Technology" (G. C. Sih and C. L. Chow, eds.), Vol. 1, p. 297. Sijthoff and Noordhof, Alphenaan den Rijn, The Netherlands, 1977.

[185] S. C. Kunz and P. W. R. Beaumont, ASTM STP 569, p. 71 (1975).

[185a] A. Rotem, NASA, in press, 1980.

[186] P. J. Worthington, paper presented, *ASTM Symp. Fatigue Fibrous Composite Mat., San Francisco, May 22–23* (1979).

[187] M. B. Kasen, R. E. Schramm, and D. T. Read, ASTM STP 636, p. 141 (1977).

[188] K. L. Reifsnider, W. W. Stinchcomb, and T. K. O'Brien, ASTM STP 636, p. 171 (1977).

[188a] G. P. Sendeckyj, and H. D. Stainaker, ASTM STP 617, 39 (1977).

[189] J. B. Romans, A. G. Sands, and J. E. Cowling, *Ind. Eng. Chem. Prod. Res. Dev.* **11**, 261 (1972).

[190] H. T. Sumsion and D. P. Williams, ASTM STP 569, p. 226 (1975).

[191] W. D. Bascom and R. L. Cottington, *J. Adhes.* **4**, 193 (1972).

[192] J. F. Mandell, *Polym. Eng. Sci.* **19**, 353 (1979).

[193] R. L. Patrick (ed.), "Structural Adhesives with Emphasis on Aerospace Applications" Vol. 4, Treatise on Adhesion and Adhesives. Dekker, New York, 1976.

[194] G. P. Anderson, S. J. Bennett, and K. L. DeVries, "Analysis and Testing of Adhesive Bonds." Academic Press, New York, 1977.

[195] J. A. Marceau, J. C. McMillan, and W. M. Scardino, *Sci. Adv. Mat. Process Eng. Ser.* **22**, 64 (1977).

[196] W. D. Freeston *et al.*, Skirts and Seals for Surface Effect Vehicles, Rep. NMAB-340. Materials Advisory Board, Commission on Sociotechnical Systems, National Academy of Sciences, 1978.

[197] S. Mostovoy and E. J. Ripling, *Polym. Sci. Technol.* **9B**, 513 (1975).

[198] J. A. Marceau, Y. Moji, and J. C. McMillan, paper presented at the *SAMPE Symp.*, *21st*, *April 6–8* (1976).

[199] W. D. Bascom, C. O. Timmons, and R. L. Jones, *J. Mat. Sci.* **10**, 1037 (1975).

[200] D. J. Wilkins, R. V. Wolff, M. Shinozuka, and E. F. Cox, ASTM STP 569, p. 307 (1975).

[201] J. A. Manson, *Mat. Sci. Eng.* **25,** 41 (1977).

[202] G. C. Hoff, *et al.* (eds.), "Chemical, Polymer, and Fiber Additives for Low Maintenance Highways," Chapter 10. Noyes Data Corp., Park Ridge, New Jersey, 1979.

[203] G. Gebauer and J. A. Manson, unpublished research, Lehigh Univ. (1974).

[204] L. E. Kuckacka, J. Fontana, and A. Romano, Concrete-Polymer Materials Development Progress Report for the Federal Highways Administration, unpublished report, Brookhaven National Laboratory for the quarter July–September 1973.

[205] D. W. Fowler, J. T. Houston, and D. R. Paul, "Polymer-Impregnated Concrete for Highway Applications," Rep. 114-1, February 1973, The Univ. of Texas at Austin, Ctr. for Highway Res.

Index to References*

* Numbers indicate the pages on which the complete references are listed. Numbers in parentheses indicate the number of references by the author on that page.

Subject Index

Materials Index